Making Sense

A Student's Guide to Research and Writing

Engineering and the Technical Sciences

Fifth Edition

Making Sense

A Student's Guide to Research and Writing

Engineering and the Technical Sciences

Margot Northey

Judi Jewinski

OXFORD

UNIVERSITY PRESS

OXFORD
UNIVERSITY PRESS

Oxford University Press is a department of the University of Oxford.
It furthers the University's objective of excellence in research, scholarship,
and education by publishing worldwide. Oxford is a registered trade mark of
Oxford University Press in the UK and in certain other countries.

Published in Canada by
Oxford University Press
8 Sampson Mews, Suite 204,
Don Mills, Ontario M3C 0H5 Canada

www.oupcanada.com

Database right Oxford University Press (maker)

First Edition published in 2005
Second Edition published in 2007
Third Edition published in 2009
Fourth Edition published in 2012

Library and Archives Canada Cataloguing in Publication
Northey, Margot, 1940-, author
Making sense : a student's guide to research and writing : engineering
and the technical sciences / Margot Northey, Judi Jewinski. -- Fifth edition.

(Making sense)
Previous edition published by Oxford University Press, 2012.
Includes index.
ISBN 978-0-19-901025-7 (paperback)

1. Technical writing. 2. Report writing. 3. English language--Rhetoric.
I. Jewinski, Judi, 1952-, author II. Title. III. Series: Making sense series

T11.N67 2015   808.06'691   C2015-904242-9

Cover images (clockwise from top left): © iStock/MiguelMalo; © iStock/alex_rext; © iStock/jewhyte

Oxford University Press is committed to our environment.
This book is printed on Forest Stewardship Council® certified paper
and comes from responsible sources.

MIX
Paper from
responsible sources
FSC® C103567

Printed and bound in Canada

1 2 3 4 — 19 18 17 16

Contents

Acknowledgements

I continue to be grateful for the collaboration of my colleagues at the University of Waterloo who have helped me improve this book for students. Included in this list are Catherine Burns, Donna Ellis, Katherine Lithgow, Wayne Loucks, Wayne Parker, Peter Roe, Gordon Stubley, and Andrew Trivett, all of whom have cheerfully shared examples and given me very good advice. Sandra Keys has been an invaluable reference for citation conventions, and I appreciate the help of software developers Stefan Jewinski and Madeleine Hart in making sure examples are timely and up to date. And the chapter on graphing and formatting has benefitted enormously from the expert suggestions of my EARTH 10 co-instructor, Brewster Conant, Jr, whose attention to detail is second to none.

My thanks, too, to all those readers from across Canada whose comments have helped shape the revised text of this fifth edition. For their hard work and sensitive suggestions, I am indebted to the editing team at Oxford University Press, especially Leah-Ann Lymer, whose patience and understanding are exemplary. Throughout this new edition, we've worked hard to emphasize the importance of meeting the readers' expectations. I do hope I continue to meet those of the readers closest to me—my ever-expanding family and the man I've been happily married to for more than 40 years.

A Note to the Student

This book has been developed for students of the engineering and technical sciences. Its purpose is to provide a framework first for conducting research and then for writing clearly and comprehensively.

Engineering is a blend of science, industry, technology, mathematics, and business; the assignments you complete as part of your studies require you to bring together theory and practice not just accurately but intelligibly and convincingly. This book shows you how to refine your research and writing skills to present ideas professionally both on paper and in person.

This fifth edition of *Making Sense in Engineering and the Technical Sciences* reflects continuing trends in technology that simplify the production of documents, although not their creation. It contains some important revisions including new material on paraphrasing and writing summaries as well as templates for developing topic sentences and paragraphs. Documentation styles have been updated, and each chapter now ends with a Chapter Checklist to help you review the main points. Writing in university or college is not fundamentally different from writing elsewhere. Yet each piece of writing has its own special purposes, and these are what determine its shape and tone. *Making Sense in Engineering and the Technical Sciences* examines both the general precepts for effective writing and the special requirements of academic work in engineering and related fields. It also points out some of the most common errors made by inexperienced writers and suggests how to avoid or correct problems. Written mostly in the form of guidelines rather than rules, this book should help you escape some common pitfalls and develop confidence through an understanding of basic principles and a mastery of sound techniques.

The structure of this book follows a standard developmental process from design to prototype to product. Above all, it is intended to be a clear, concise, and readable guide that helps you do well in all your courses—and long afterwards in your professional career.

Writing and Thinking

> **Chapter Objectives**
> - Planning and executing professional writing projects
> - Meeting readers' expectations of content, organization, and style

Introduction

To produce clear writing you first need to have done some clear thinking, and clear thinking can't be hurried. It follows that the most important step you can take before beginning to write is to leave yourself enough time to get organized.

Remember that writing is very much a process of recording ideas for yourself in notes and then of preparing to share those ideas with others in some written form. Whether you are communicating formally or informally, there are three steps to this process: the planning, the drafting, and the refining. The planning stage calls for you to assess the various requirements of the job at hand, such as the format, the purpose, the audience, and the approach. When the time you take researching and preparing to write is well spent, the final two stages—the drafting and refining—are a lot easier.

Initial Strategies

Developing a project requires you to make choices about what ideas you want to present and how you want to present them. The decisions become easier to make with practice, but you still always have choices.

With every project you undertake, it's sound strategy to ask yourself two basic questions:

- What is the purpose of this?
- Who is going to read it?

Your first reaction may be, "Well, I'm writing for my lab instructor to fulfill a course requirement," but that's not useful. The people assessing your work expect you to recognize the demands of the context, to demonstrate that you can imagine—and write to suit—the professional setting for your work, even though you might not graduate for another three or four years.

Think about the purpose

On the job, you work as part of a team to present solutions to clients or to conduct independent research. In school, your courses are designed to prepare you for such responsibilities. Coursework is thus intended to confirm that you understand the concepts presented in class and have learned to apply them. Depending upon the assignment, your purpose may be one or more of the following:

- to record and present data from tests and experiments that you have performed
- to show that you understand and can apply certain principles, concepts, or theories
- to show that you can do independent research
- to confirm your ability to solve problems and share convincing solutions
- to demonstrate that you can think critically
- to demonstrate that you can think creatively
- to confirm your professionalism
- to show your ability to synthesize ideas and present evidence
- to record stages in the development and completion of a project
- to confirm what you know and show that you can apply it to specific situations.

A lab assignment in which you must show that you have understood and confirmed a hypothesis calls for an approach that's different from the one you take when you write a proposal to solve a problem. In the first, your approach is straightforward, with an emphasis on the accurate recording and reporting of a process. In the second, you must provide convincing support for your recommendations, and you achieve that goal by anticipating and overcoming potential objections.

You may also want to bear in mind the weight of an assignment within the context of the overall marking scheme. A project worth 50 per cent or more of your final grade merits a significant investment of time and effort. No matter how heavy your workload, never leave such assignments to the last minute. Group projects, too, require extensive commitment, with the complication of making sure everyone shares the work equally.

While your primary purpose will differ from one type of writing to the next, your secondary purpose will always be the same: to confirm that you not only understand what's expected but are capable of demonstrating it.

Think about the audience

Thinking about your reader means recognizing that there may be others besides an instructor involved in evaluating your assignment. Eventually, there may be both primary and secondary readers. This will almost certainly be the case when you are writing in a professional context, particularly if you are completing an apprenticeship or co-op placement as part of your studies.

- Your primary readers (your professors or your supervisors) are experts in the field, and you can expect them to be knowledgeable about the subject.
- A secondary reader, on the other hand, may not even be an engineer, so you also have to consider how much special explanation might be necessary or relevant.

Writing for multiple readers is a skill all professionals must develop. Start by imagining how you might describe a concept like desalination to a lab technician, then to your aunt or to a friend who's in economics. Each reader has different needs and expectations, and no reader wants to waste time working through information he or she doesn't need. Ask yourself what each one of your readers *wants* to know and what each *has to* know. Once you know who you are writing for and why, you can address those needs directly in a cover letter or executive summary that accompanies your document. In doing so, you can focus your message not just for an expert but also for a non-technical reader who wants only to know the results.

Think about the research question

Before you set out to explore any subject, you need to have a pretty clear idea of what you think you are looking for. If you take the time to put your thoughts into a sentence, you will find it falling somewhere between a *statement of purpose* and a *thesis statement*. This sentence, or *research question*, will become the controlling idea for your research. After you've done the background work, you can come back to this sentence and adjust it for your results. (Of course, if you begin with a hypothesis that your results eventually disprove, you must nevertheless present these findings fairly and then revisit your hypothesis for your next project.) The research question is the central statement in the writing that is meant to share your findings with others.

You may recognize the pattern: the development of this statement comes at the beginning of the approach known as the *scientific method*, which involves the following steps:

1. formulating and delineating the problem
2. thoroughly reviewing related literature
3. developing a theoretical framework
4. formulating hypotheses
5. selecting a research design
6. specifying the object or population for study
7. developing a plan for collecting data
8. conducting a pilot study and making revisions
9. selecting the sample
10. collecting the data
11. preparing the data for analysis
12. analyzing the data
13. interpreting the results
14. sharing the findings with others

The last step of this process, the formal writing up of results, will highlight the controlling idea in the introduction, perhaps phrased as a question, and confirm the answer in the conclusion. Take the time to frame your thesis carefully at the start of your project by asking the right question: *what? how? why? where? when?* or *who?* Note how each response suggests a way of developing and organizing the content. The following examples illustrate this progress from question to answer to structure:

- *What* are the best turbo-coding algorithms?
  - *The best ones are ___ and ___.*
- *How* much fertilizer will produce optimum crop yields of barley?
  - *The optimal amount is ___.*
- *Why* do rechargeable lithium-ion cells store energy best?
  - *Lithium-ion cells are the best for three reasons.*

- *Where* is the best location for emergency water supplies in Red Deer?
  - *The two best locations for emergency water are ___ and ___.*
- *When* is the critical maintenance time point for the King Street Bridge?
  - *The King Street Bridge needs maintenance work completed by mid-2016.*
- *Who* stands to benefit most from applications of nanotechnology?
  - *Nanotechnology holds particular promise for two groups of people.*

The answers, which take more or less research to come up with, are nevertheless limited and precise. Use this question-and-answer structure to provide a focal point both for you in your project and, more usefully, for your eventual reader.

Think about the structure

A number of blueprints exist for the formal presentation of written material. These are all variations on the basic introduction-body-conclusion structure you've been working with since your very first composition. What you may not have known is that this same arrangement reflects the process of reasoning common to all empirical sciences. One of the most common applications of the scientific method is the technical report, which objectively presents the results of experiments and tests (see Chapter 4). Of course, the analysis done in such writing is largely theoretical. This complaint of work done in academic contexts (that it's far from what many refer to as the "real world") is a common one. But theoretical analysis is an important first step. The practical writing you learn to do on the job refines the principles of academic writing to develop discussions that people can read and use as a basis for making sensible decisions. This practical application is generally associated with problem solving.

The step-by-step system for arriving at results can also guide you as you develop a plan for writing. For one thing, it provides a means of sorting, or classifying, information into concrete observations and the generalizations that can be made about them.

- If you come to a conclusion based on what you have observed, it's called *inductive* reasoning, or *induction*.
- When you begin with a hypothesis and proceed to test it according to evidence you uncover, it's called *deductive* reasoning, or *deduction*.

Archimedes' discovery of the laws of buoyancy (said to have occurred when Archimedes stepped into a full bath, causing it to overflow) is the model for this organizational approach, represented in Fig. 1.1.

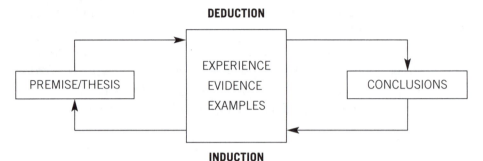

Figure 1.1 Inductive and deductive reasoning

Whether you start with the evidence and come to a conclusion about it, or start with a hypothesis and discover the evidence to support or reject it, this pattern blends general and specific content. Take advantage of its value as a pattern for any writing you do—in a course or on the job.

Here's the basic model to follow:

$$\boxed{\text{statement}} \;+\; \boxed{\text{specifics}} \;+\; \boxed{\text{summary}}$$

For every main point, some kind of development is always necessary. Whether it's a one-paragraph answer to a question on a test or it's a section in an extensive project, your organization imitates this fundamental pattern. You will find this model especially useful as you develop skills for writing summaries of your own work and of the work of others (see Chapter 3). In fact, it enables you, both methodically and systematically, to take advantage of complementary organizational patterns.

There are generally two basic patterns:

- one organized by *time*, such as you might find in a technical report, and
- one organized by *priority*, such as you might find in a proposal or in a paper addressing a problem.

Once you have done the background work, you approach a writing project the same way you'd tackle any problem-solving or design exercise—by selecting the most promising organization for each section in light of what you want to say about your hypothesis or research question.

The following are traditional models based on patterns of enumeration. Note how each general statement (**G**) anticipates the *several* points that follow.

These are listed point by point in a predictable order of importance: *first*, *second*, *third*, and so on, to *finally*. (When you adapt such models to your own context, of course, you not only fill in the missing topic but also replace the word *several* with an appropriate number.)

- **Process** (time order). This model shows how something works or has worked. Whether used for a list of instructions or a record of observations, it subdivides material into a series of stages or steps:

  (**G**) There are *several* steps in the development of _____.

- **Cause/effect** (time order). This model is used to represent an applied process. Beyond merely showing the order of events, it emphasizes their interrelationship. When using this model, be careful to distinguish between a direct cause (*A produces B*), a contributing cause (*A helps produce B*), a condition (*If A, then B*), and a coincidence (*Both A and B exist*).

  (**G**) _____ has *several* causes.

  (**G**) _____ has *several* effects.

- **Description** (space order). This model identifies the composition of something, from a piece of equipment to a site, by listing distinguishing details and characteristics. The arrangement follows a logical progression, say from left to right, and is usually supplemented by a diagram or map. This pattern is typical of engineering specifications.

  (**G**) _____ has *several* features.

  (**G**) _____ has *several* constraints.

- **Classification** (space order). This model is a method of dividing something into components according to a principle of selection which you define.

  (**G**) There are *several* types of _____.

  (**G**) There are *several* criteria for _____.

Smart phones, for example, may be classified according to manufacturer, applications, security, cost, or practicality—and each subdivision might involve different members. It's also possible to classify people. Students, for example, may be full- or part-time, regular or co-op, engineers or other. There are two strict rules to follow when classifying groups:

1. You must be able to account for all members of a class. If any are left over, you must adjust your categories or add to them.
2. You can divide categories into two or more subcategories as long as there are significant differences within one grouping. The standard classification of rocks into three types—metamorphic, igneous, and sedimentary—allows for an examination of subcategories too.

- **Comparison** (space order). This is a way of considering two members of the same class with respect to both similarities and differences. Although it is possible simply to follow descriptive order, characterizing the first item and then the second, comparisons are more effective if you do a point-by-point examination, considering the two items together, feature by feature.

  (**G**) Although they share certain characteristics, A is different from B.

  This organizational pattern is common in conclusions and recommendations, where you present your alternative as *better than* others.

- **Contrast** (space order). This is a comparison that considers only the differences.

  (**G**) A and B are different.

- **Problem/solution** (time order). This model generally depends on cause/effect order to identify something as a problem to be solved, then on process order to express the methodology. Comparison offers a useful structure if more than one solution is possible.

As you develop an outline for your writing, you can save yourself organizational time by identifying how these patterns can govern your overall arrangement of material, paragraph by paragraph and section by section.

Think about the format

Take time to think about what the final product involves in terms of appearance, layout, length, and style. Attention to format means understanding your reader's expectations as well as those of the discipline. What works for an engineering project won't necessarily work elsewhere. If you were taking an elective course in classics, for example, you would have to become familiar with and follow the conventions of the Modern Languages Association (see Chapter 9).

Before you begin, then, consider what it is you are preparing. Are you writing product specifications, a technical report, a letter, a position paper, an article, a progress report, an analysis, a lab report, a proposal, or an abstract? Examine examples and models carefully, including those in this handbook, so that you understand the conventions. Understanding expectations is the first step to meeting them.

Think about the length

A very basic rule in professional writing is that what's written should be "no longer than it needs to be." Before you even start the planning phase, recognize potential restrictions on the length of the assignment. If you have been asked to develop your own project, trust someone with experience to act as your guide. Consult with your instructor, professor, or supervisor, and refer to examples of similar assignments available in the library, in the campus writing centre, or on course websites to give you an idea of how to limit your focus. Of course, there are often strict length requirements associated with assignments. (An abstract must be kept to fewer than 250 words, a cover letter kept to a page, and so on.)

Think about the tone

Tone refers to a writer's use of language. The words friends use together may be casual and include a lot of slang, but the words used in a professional setting, whether at school or on the job, are usually formal and specific. The language in a piece of writing reflects the familiarity between the writer and the audience. Two friends are close; students and professors are less so. When you e-mail friends or colleagues, a casual tone is both natural and appropriate. An e-mail to a professor or a personnel committee, however, will take a much more formal tone.

As Figure 1.2 points out, just how formal you need to be depends both on the assignment and on the instructions you have been given. In some cases— for example, if your fluid mechanics instructor expects you to turn in your lab notes—you may be able to write in point form. When you write instructions for someone, you can comfortably refer to the reader as *you*. This style of address is clearly more personal than the conventional, more formal tone of most projects, reports, and papers. Choosing the appropriate register and maintaining a consistent approach throughout your work is one of your most important responsibilities as a writer. Begin to determine what's appropriate by identifying where your writing fits on the scale in Figure 1.2.

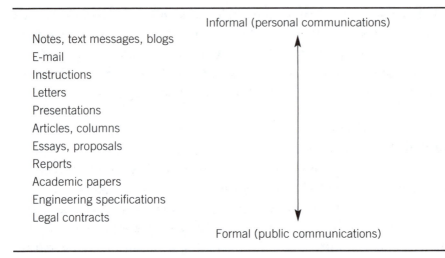

Figure 1.2 The scale of formality

Depending on the type of assignment, your purpose, and your audience, your writing will contain features that move it one way or the other along this scale of formality. For example, the tone of this handbook is fairly informal, because its purpose is instructional. We address you personally, we talk to you rather than write at you, and we keep the words simple. Most engineering contexts, however, call for more businesslike formality. Thus you will want to recognize—and minimize—features of your style and tone that make your writing too casual for the academic or professional situation. At the same time, you need to be careful of sounding unnaturally formal, with language that belongs in another century. The following features make your writing less formal. Above all, you want to be sure to match the tone to the situation.

Colloquial language or slang

The language you use every day seems natural, but it belongs to casual contexts like e-mails. Using informal language in the wrong context suggests that you don't recognize or appreciate the important distance between yourself and your reader. Consider the language you choose to be like the clothes you wear. Do you wear the same clothing to an outing with friends that you would wear to an interview with a prospective employer? Indeed, if you use words like "This needs to get fixed" instead of "This needs repairing," your language seems to be wearing denims rather than a suit.

On the other hand, you don't want your language to appear overdressed (*exaggerated* is perhaps a better way to put it). A tuxedo is fine attire for a wedding, but inappropriate for the office. Similarly, it is more fitting to write "the device was repaired" than to write the somewhat artificial "the contrivance underwent reparations." Make sure that the language you use in writing shows appropriate respect for the situation and for conventions. Don't use slang, but don't develop an overblown vocabulary either.

Short forms and abbreviations

Even though it's quick and easy to rely on popular instant-messaging abbreviations (*cy, btw?*), they are never appropriate in the writing you do for someone you hope to impress, especially someone who may be giving you a job or a grade. Even if you are e-mailing a teaching assistant a question about homework, use standard unabbreviated language that proves you are not in too much of a hurry to be polite.

Contractions

Contractions such as *can't, it's, won't,* and *we'll* are not suitable for formal academic writing, although they are common and natural in letters or, for that matter, any writing that's intended to be read aloud, like a paper designed for an in-class presentation. If you want to use contractions, remember that they make writing sound chatty and informal. If that's your purpose, then the contraction is justified. (Note our regular use of contractions like *that's* to address you directly.) Remember, however, that some readers don't accept contractions. It's a good idea to spell things out in writing for people you haven't met.

Point form

Bullets simplify lists for readers. When you use them, follow conventions of presentation to distinguish between formal and informal types. A complete sentence ending with a colon (:) precedes formal bullets, which have their own conventions:

1. They use punctuated numbers instead of dots or dashes or diamonds.
2. They include appropriate punctuation at the end of each entry.
3. They are grammatically complete and use parallel wording (see Chapters 14 and 15 for conventions of grammar and punctuation).

Informal bullets use point form without punctuation, except perhaps for periods if each one is a complete sentence. Choose the bullet format to suit your situation.

Personal pronouns

In personal communication, whether spoken or written, it's natural to use personal pronouns such as *I*, *you*, or *we*. In fact, it's impossible to address anyone directly without them. The objective nature of scientific and technical writing, however, emphasizes evidence, results, and recommendations rather than the people responsible for them. In fact, many readers object that a dependence on *I* reveals a lack of certainty or conviction, definitely not an impression you want to convey:

> ✗ I feel the concept behind the application is sound.

> ✗ In my opinion, the concept behind the application is sound.

> ✓ The concept behind the application is sound.

At the same time, there needs to be a way to avoid reconstructing sentences with awkward passive constructions (for example, "It can reasonably be concluded, based on the evidence that has been presented, that . . ."). Technical writing does permit the substitution of *we* for *I*, especially when reporting the results of a team investigation. Substituting *the team* for the sometimes controversial first-person pronouns is another option:

> ✗ It was determined that the prototype that had been developed by the team would be too expensive to produce.

> ✓ We discovered that the prototype would be too expensive to produce.

> ✓ The team recommends abandoning the prototype, for it would be too expensive to produce.

Keep *I* and *you* for less formal situations, like discussion posts, blogs, letters, and oral presentations. (A hint: when you do use *I*, it is less obtrusive if you place it in the middle of the sentence, rather than at the beginning—a point to bear in mind for cover letters, especially the one accompanying your resumé.)

Good writers take the time to write well—either formally or informally. More importantly, they know what a situation calls for and keep both their tone and register consistent—whatever approach they've decided is appropriate.

You can learn a lot about appropriateness from the people with whom you are studying. The best writers keep to a level and approach that is unobtrusive—no one notices the language they use because it is so fitting. Refer to Chapter 13 for further help keeping your language both suitable and readable, whether your context and the audience call for you to be formal, informal, or somewhere in between.

Guidelines for Writing

Whenever you embark upon a writing project, keep the following guidelines in mind:

- Be clear about your subject and your purpose, what it is you expect to achieve.
- Think about your audience, the reader or readers of your writing.
- State your purpose clearly.
- Define your terms.
- Make sure you are accurate in all of your statements, in your analysis and presentation of data, and in your documentation of sources.
- Arrange your material logically.
- Include only relevant material. Don't pad your writing to achieve a certain length.
- Draw conclusions that are based on the evidence.
- Be simple, clear, and consistent in expressing your ideas.
- Choose and maintain an appropriate tone and level of formality throughout.
- Allow yourself plenty of time to work on drafts before completing the final copy of major and minor projects.
- Edit and proofread carefully.

Consider the flowchart on the following page (Fig. 1.3) a framework for the various stages of a writing project—from conception to submission. At a number of points, you revisit stages already completed, modifying as necessary, so that you end up with a professional piece of work that will satisfy everyone's expectations.

The chapters that follow address each of these guidelines in detail.

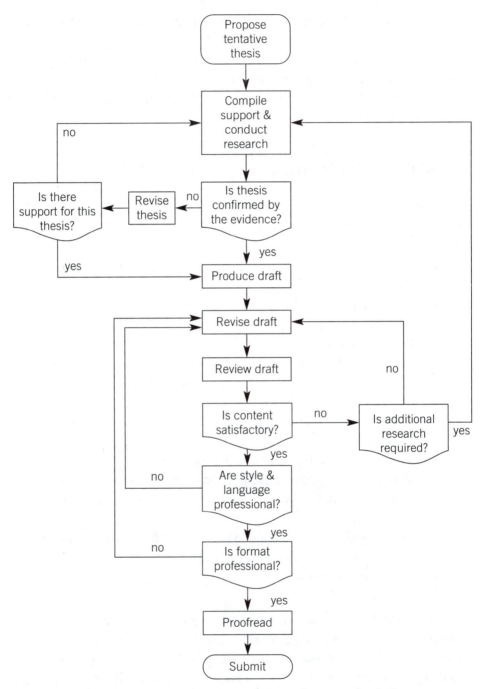

Figure 1.3 The writing process: From conception to submission

Chapter Checklist

- ☐ Identify your purpose for writing, and consider the needs and expectations of your audience.
- ☐ Organize your project according to recognized patterns for introducing and developing ideas.
- ☐ Situate the format and tone of the writing according to its position on the scale of formality.
- ☐ Prepare to work through the writing process from conception to submission.

Keeping Notes and Doing Research

2

Chapter Objectives

- Evaluating sources critically
- Taking effective notes
- Avoiding accusations of plagiarism

Introduction

One of the most important habits you should develop as an engineer is the practice of keeping extensive and accurate notes. Carefully cross-referenced files of articles, references, data, ideas (inspired or not!), sketches, and so on provide a permanent record you add to throughout your career. Ultimately, your notes and files establish a foundation for everything you write. It's important, therefore, that they be thorough and exact.

Develop the habit of carrying a notebook at all times. Whether you use a smart phone, a tablet, or a traditional notebook (preferably one with sewn-in pages), try not to delete something you aren't happy with. All entries represent a valuable record even if they seem unsatisfactory at the time. To delete material from handwritten notes, use a simple X so that the contents can still be read. Your engineer's notes, with every entry dated, help you trace the chronology of your work on an enterprise and, when corroborated by an independent third party, even serve as proof of authenticity in case of legal disputes. What better proof of ownership of intellectual property? Indeed, if you are working on a sensitive project, it is good to have an independent person corroborate your work. Electronic records make it easy to confirm chronologies, so it's essential to keep back-up files.

In addition to the records in your personal files, you will find good note-taking skills invaluable in three common school situations: in the classroom, in the lab, and in the library.

Lecture Notes

Learning to recognize the classroom manner of your professors and lecturers will help you take the most efficient notes in their classes. Some instructors rely heavily on textbooks, class notes (either bound or downloadable from the course website), slides, and handouts. Add your own comments to these directly, highlighting important points and expanding on them in margin notes throughout each lecture.

Other instructors assign a textbook as supplemental reading and spend class time on applications. In these classes, be prepared to take more substantial notes: record everything the professor or lecturer writes, keep track of page references and URLs, use headings and subheadings as a chronological record of what was covered in class, and copy all equations, calculations, and formulas accurately. If your instructor relies on slides, be sure to download these as soon as they are posted on the course website, adding your notes from class to the file or to a hard copy.

Whatever the lecture style, take advantage of the following time-honoured techniques for making the most of your classes:

- For each course you take, keep a binder for handouts and create an electronic folder to include your own notes as well as copies of slides and supplemental readings. If there's a learning management system attached to your course, familiarize yourself with its contents and attach your notes to each topic.
- Be prepared for class. Note-taking is much easier if you know beforehand what the topics of discussion will be. Keep track of the course outline so that you know what content to expect from class to class. Do assigned problems or readings before class, and review previous notes too.
- Start a new file for each week of class using the date as the file name. Take note of any organizational overview given in class, recording headings and summarizing cases. SMS lingo can help you develop your own shorthand, building on the following abbreviations:

=	is, are, equals	**w**	with
≠	does not equal / differs from	**w/o**	without
≈	is like, approximately	**+**	and
⇒	leads to, produces	**/**	or

Ø	nothing	¶	new paragraph
>	greater than	!*°	important
<	less than	? or Q	query, question
⇑	increases	A	answer
⇓	decreases	±	plus or minus
cf	compare with, see also	@	at
∵	since	P	page
∴	therefore	x	times
∀	in all cases, for all	Iff	if and only if

- Create your own acronyms or short forms for common terms or topics.
- If you have software that allows you to record the lecture while you take notes, you'll soon appreciate being able to replay the sections that you want to work on.
- Whether you take notes on a tablet or on paper, leave a wide margin on the right-hand side so that you have room to add summaries or cross-references when you are studying for exams.
- Highlight important points, or mark them with an asterisk. Put a question mark next to points to reconsider and clarify after class—either with classmates or with the course instructor.
- Use columns and tables to record comparisons. Take advantage of the abbreviations Q and A (question/answer), A and D (advantages/disadvantages), C and E (cause/effect), or P and S (problem/solution) to identify relationships.
- Record all key terms, including names, numbers, and nomenclature. Write out all definitions.
- Take pictures or otherwise make copies of all diagrams, sketches, charts, and graphs.
- Keep a master calendar on which you track all assignments and due dates.
- Take time, as soon as you can after class, to review and refine your notes, being sure to set down questions that might show up on an exam. If you have recorded references to external material (the textbook, websites, course notes, etc.), read these now and add summaries as necessary.

Taking notes electronically

If you have note-taking software, take advantage of all the shortcuts it provides, including the ability to record audio, take videos, and download course material to annotate. Be aware that most professors prefer tablets to laptops. If you are allowed to use a laptop, be aware that you may disturb students around you as well as the course instructor. Here are some guidelines for using one courteously:

- Sit at the sides or back of the classroom to provide minimum distraction for your professor and classmates.
- Have your laptop turned on before class begins, and mute the volume.
- Use your laptop exclusively for taking lecture notes.
- Wait until the end of class to put away your laptop.

Sharing class notes

You will naturally develop close relationships with classmates during the term. Whether or not you establish study groups, make sure you develop reliable contacts early on in case one of you has to miss a class. Don't abuse a friend's willingness to share notes, and always be prepared to return the favour. Remember that it's considered quite unprofessional and inappropriate to post a general request for notes on the course discussion board. If you have to borrow notes, don't just copy them. Instead, take the time to work through them as if you had been to class.

Lab Notes

When you take notes in class, you must adjust to the pace of your instructor. In the lab, however, you control both the process and the timing. You determine how much time you need to take to keep accurate and complete records.

Success lies in your thorough preparation for the experiment and any pre-lab work, so that you are well organized and ready to record results.

- Complete all required pre-lab exercises so that you are well prepared. Review and record the context, the methodology, and any apparatus for the experiment. Many professors require that you note the specifics of the equipment you are using (brand names and serial numbers, for example), in case disparate results later need to be justified.
- As necessary, prepare data grids, spreadsheets, or other templates to be filled in during the experiment. Whatever you can do to simplify the recording of data during the lab will make the job of analyzing it later on easier. It will also minimize the risk of misreading the data.

- Make sure you include the date and the name or title of the experiment at the beginning of each entry.
- Be systematic and thorough in recording calculations and equations. Be sure to identify all terms.
- Supplement your notes with video clips and photos throughout the experiment.
- Keep your conclusions simple and precise. Before you submit your lab report, make sure that you have answered all the questions.

Although computers have made life easier, many professors still require students to purchase bound notebooks to use to record all lab work done in a course. If you have such a lab instructor, here are guidelines for meeting his or her expectations:

- The best lab books have hard covers and sewn-in pages of graph paper. Number the pages and reserve three or four sheets at the front so that you can later add a dated title page and general table of contents (with page numbers for each lab the book includes). As you work page by page through the book, put each day's date at the top of the page, just below the page number.
- Write in ink. If you are working with a lab partner, consider having each of you write in different coloured ink so that your contributions can be distinguished easily.
- If your lab calls for computer manipulation of data, paste the printout carefully into the lab book. Write the folder name and file name directly into your book so that you won't waste time finding it later. Some students record related files on CDs to keep in pockets glued to the inside covers of their notebooks too.

Remember that lab work is a process. By recording it carefully and chronologically, you make it easy for a reader to retrace your steps, follow your reasoning, and even duplicate your results.

Research Notes

Research is done for many reasons: to keep up to date with scientific trends, to verify what other work is being done in your field, and to spark or develop ideas for projects. When you are not in class or in the lab, your life as a student likely puts you in front of a computer. This is a good place to start identifying materials for your research, and you might as well be systematic about your approach. Use your note-taking skills to keep track of your sources as well as to record the information you find.

Explore library resources

The importance of getting to know your way around the resources available through your institution's library can't be stressed enough. It's worthwhile to take part in an orientation seminar at the beginning of term and to get to know the people at the information desk. Librarians are glad to show you how to access the bibliographies, indexes, databases, journal abstracts, and other reference tools for your particular field. Once you are familiar with the basics, you are well prepared to take advantage of available resources, including both electronic and print materials.

In addition to on-site resources, libraries offer access to a wealth of online services that are accessible from anywhere. The homepage of your college or university library is an ideal starting point. Take advantage of online tutorials to become more familiar with the possibilities these resources offer.

Take the time to look up works listed in the bibliographies and references of your course textbooks and journal articles referred to in class. You might come across related material that would be difficult to find otherwise.

Online catalogues list all the holdings at your library, including books, videos, microforms, archived materials, and print journals. A search by subject or keyword will give you a list of relevant sources. Don't forget that you can also search the national libraries of Canada and the United States, which are repositories for everything published in these countries.

Databases simplify your search for information because they make millions of journal articles and electronic resources available in a single place. Libraries subscribe to online services like e-books, e-journals, e-data, and electronic encyclopedias. A single search gives you access to opportunities for further investigation.

To conduct a database search, simply go to your library website and follow the link to resources for research. You can then search by keyword, subject, author, or title. There are options for narrowing the search if you prefer, for example by restricting it to specific databases or specific journals. Your search results will be given in the form of a list of articles on your subject, including the title, author, publication information, and an abstract. You will likely be given the option of downloading a file of the full text of the article, which allows you to read the material online as well as print, save, or e-mail it.

If you are having trouble finding just what you want, don't hesitate to ask for help from a librarian. These professionals are intimately familiar with the various means for unearthing information and can give you guidance and direction or walk you through a search. Most college and university libraries offer information services with names like *Ask a Librarian*. Without leaving your computer, you can connect with a librarian—either by going to an online chat room (or telephoning) during designated hours or by sending e-mail queries. If you take advantage of such help while preparing a major project, don't forget to thank your librarian in your acknowledgements.

Explore Internet resources

In addition to the online services provided by your library, it is easy to access the huge volume of information available on the Internet from billions of different websites. Be careful, though. You need to invest time and energy to make sure that the information you get is reliable and accurate. When an instant search yields over a million hits, you need to be able to assess the validity of what you get and find a way of narrowing the search.

Evaluating online sources

It is crucial to distinguish between academic websites, which are peer reviewed and tend to be reliable sources of information, and the vast majority of websites which do not have editorial boards and publish material that has not undergone any review process. Remember that anyone can upload material onto the Internet, regardless of his or her credentials (or lack thereof), as the profusion of blogs attests. Before trusting any material, make sure that the author or publisher has the necessary authority to lend credibility to the site.

Here are some tips for evaluating material you find on the Internet:

- Look for a statement identifying the site host, the author's qualifications, and contact information. Many sites are affiliated with an educational or professional organization, and sites hosted by an individual should provide you with specifics about that person's qualifications. It's possible to identify domain and IP owner information behind any website by performing a search at www.who.is.
- Check the URL. Pay special attention to the domain name, which may help you assess the objectivity of the site. Be aware of the differences between and the motivation of sales-oriented enterprises (ending in *.com* in North America and *.co.uk* in the UK, also *.biz* and *.net*) and

not-for-profit or public-service organizations (ending in *.org*). Although the main purpose of a *.com* domain is self-advertising and an *.org* domain may have an obvious agenda, any one of them may still provide reliable information and valuable links to other sources.

- Recognize that not all commercial sites have obvious domain names. Because every country has its own two letter abbreviation (for example, *.ca* is for Canada, *.de* is for Germany, and *.mx* is for Mexico), you may link to information sponsored by businesses anywhere in the world. Note also the domain names for government sites (*.gc.ca* in Canada and *.gov* in the United States) and academic institutions (*.edu* for American ones). And just because the site is that of a college or university doesn't guarantee the quality of the material you find. When in doubt, ask an instructor or your school librarian. Your increasing good judgment and experience will help you decide whether material on any site is trustworthy.

- Determine the currency of the site. There should be a clear indication of the date when the material was written, the date when it was published, and the date when it was last revised. Be skeptical if the date is not recent or is missing altogether. There should be a contact e-mail address for questions. If not, be suspicious. If there are links to other sites, confirm that they work. A site with broken links is not current. Check thoroughly before you base research on information that is either out of date or simply wrong.

- Evaluate the accuracy of the information by checking facts and figures with other sources. Data published on the site should be documented in citations or a bibliography, and research methods should be explained. If there are obvious errors, do not trust the source.

- Take special care with *wiki* sites, where information is revised frequently and anyone can contribute or edit. The most popular of these, by far, is Wikipedia (wikipedia.org), which has versions in all the major world languages. Its professional look can be deceptive: errors are always being discovered, corrected, and re-corrected. Still, there's no denying the instant convenience of Wikipedia's extensive discussions. Before you include them, however, it makes sense to check their acceptability with your instructor. Often, you'll hear that the references in Wikipedia are simply a good starting point for expanded research. Remember always to validate the information elsewhere. If you use information from Wikipedia, you must cite it appropriately (see Chapter 9), but if it's your only source, you can expect criticism from your professor that you should have been more thorough in your research.

- Be wary of blogs. There are millions of them, from personal online diaries to corporate and organizational blogs like those on college or university websites that give visitors a glimpse of student life. There is value in the networking possibilities of such social sites as Twitter, Facebook, and LinkedIn, but the expression of personal opinion, even if you agree with it, is not valid research. Assess the overall quality of the site as indicated by its level of correctness and writing style. Sites with spelling errors or an inappropriately informal style lack credibility and are not to be counted on.

These are the same issues that you should consider with respect to anything you read. Developing a critical eye is just one more step to becoming a professional.

Systemize your research

More important, perhaps, than finding research material is taking notes from it that are comprehensive, dependable, and easy to use. With time you will develop your own best method, but for a start you might try creating electronic folders for each project. With this system you save the information from each source as a separate file. If your school's learning management system offers the possibility of an e-portfolio, you can keep your work throughout your education and see your progress clearly.

Use a table format with two columns. Reserve the left-hand column for source material, and put your own comments on the right-hand side. Use quotation marks for all material cut and pasted from outside sources, and record all URLs and DOIs. Unless you keep careful track of material taken from elsewhere, it can become impossible to tell where you found each piece of information. Don't risk accusations of plagiarism because you have cut and pasted too enthusiastically.

Whatever method you follow, remember that exact records are essential for proper references. The following are some guidelines for taking good notes during research:

- For every source, start with an entry that includes all bibliographical details eventually required for your reference list:
  - the name of the author(s)
  - the full title of the source
  - the place and date of publication

- the journal volume and issue numbers
- as well as the relevant page numbers.
- If it's material from a library, be sure to record its call number so that you can locate it again easily if needed.
- If it's an electronic journal, record the DOI (digital object identifier).
- If it's an online source, keep track of the complete URL, the date you accessed the material, and the date the website was last updated.

Nothing is more frustrating than wanting to use a piece of information in a paper only to find that you can't quite remember where it came from. But if you don't know the source, you cannot use the reference. To avoid last-minute panic, be thorough the first time. Here are some guidelines:

- Before you begin, confirm the citation format your instructor expects of your assignment. Then record bibliographic details following that format. When you're preparing your list of references later on, you can copy and paste in the sources accurately.

  If you are studying at a college or university, your library likely includes a free link to RefWorks, a time-saving web-based tool for managing citations and bibliographies. You can easily establish a personal account (or a joint account if you are working on a group project) and create your own database of bibliographic information for a project, importing citations and references from electronic sources anywhere. RefWorks is not the only web-based citation manager available (others include Bibus, EndNote, and Zotero), but it's the one to which most Canadian schools subscribe.

- Be sure to use quotation marks to indicate that you have copied material directly from a source. Do the same for equations. If you make it a habit to highlight or use a different font colour for everything you have copied from an outside source, you minimize the risk of forgetting to acknowledge borrowed material.
- Include page numbers for every reference, even if you paraphrase or summarize the idea rather than copy it word for word. If you are referring to a website, use paragraph or section numbers if these are available. Always copy the URL exactly.
- Keep all your notes organized carefully in folders, subfolders, and files. E-mailing yourself a copy or keeping one on a USB key is insurance that you will have access to your material as long as you need it.

Avoiding Plagiarism

When you present something as your own work, you guarantee that you are its author, except for material you have specifically identified as coming from elsewhere. If you haven't cited your sources fully and completely, or if you have inadvertently cut and pasted someone else's words into your own text without quotation marks or acknowledgement, you could be accused of plagiarism.

Plagiarism is a form of theft. As with other offences, ignorance is no excuse. Within academic institutions, penalties for plagiarism range from a grade of zero to expulsion, so it makes sense to familiarize yourself with your school's academic integrity policy. It is not worth jeopardizing a career just because you have neglected to give credit where credit is due.

It's considered plagiarism if you do one or more of the following:

- borrow from someone else's material without acknowledgement, especially when you don't use quotation marks
- paraphrase someone else's material without acknowledgement
- present someone else's ideas or examples as if they were your own
- follow the general organization or overall plan of another source, even if your subject is different.

Since research papers and projects depend on support from materials created by various sources and experts, you add to your own credibility and reputation for thoroughness when you provide support from experts in the field. A thoroughly documented paper including accurately cited material shows skill, selectivity, and thoroughness—so it makes sense to be complete and specific about material used to back up and document ideas, arguments, proposals, and conclusions.

The most obvious type of plagiarism is the deliberate misrepresentation of a source as your own. Consider the following excerpt (with underlining added for emphasis):

In designing, the two processes that must be understood are *system analysis* and *system synthesis*. Designing is a process by which new systems are made; these new systems may incorporate many components and subsystems, and it is the object of the designer to produce the best system for certain missions. Therefore, he or she must be able to analyze the operation of the components and the overall system; he or she must also be able to devise or synthesize systems of given components. *Analysis* is the <u>process of breaking down the system into parts</u> and discovering whether or not it will fulfill a mission. *Synthesis,* on the other hand, is the <u>process of building up parts into an organized whole</u> that can fulfill a mission.

One student's summary of design process includes the following passage. The underlined formatting indicates the parts that are plagiarized—exact phrasing is taken from the original with no acknowledgement given.

<u>In designing,</u> <u>new systems are made</u> by incorporating <u>many components and subsystems</u> relying on *analysis* or *synthesis,* or both. <u>The object of every designer is to produce the best system. Therefore, he or she must be able to analyze</u> the workings of the parts in relation to the whole. At the same time, <u>he or she must also be able</u> to put them together. By relying on *analysis,* <u>the process of breaking down the system into parts,</u> and *synthesis,* <u>the process of building up parts into an organized whole,</u> the designer can produce something functional.

Obviously, it is plagiarism to take selections, whether sentences or simple phrases, from a source and include them in your own text pretending that they are your own. But it is also plagiarism if you use only a phrase or paraphrase of the original without proper documentation. In the sample passage above, the scrupulous distinction between *analysis,* "the breaking down [of] the system into parts," and *synthesis,* "the building up of parts into an organized whole," should be appropriately acknowledged by quotation marks. If you left them out, even if you credited the original source in an endnote, you would still be faulted for plagiarism. The rule is simple: if the words or the points aren't yours, identify them with quotation marks and a citation note.

If you present figures or statistics whether in the text or in a table, you do not put quotation marks around them, but you must still document the source. Documentation is also required for any graphics in a format that you did not create yourself.

There is a third form of plagiarism that may not seem as blatant but is as much an academic offence as word-for-word copying. Restating ideas in your own words and phrases without acknowledgement is not a solution. You might think, for example, that you could generalize about the design process cited above by saying the following: *Analysis is easily distinguished from synthesis because the former concentrates on breaking down a system while the latter builds it up*. After all, you haven't actually used phrases from the original, and the design concept is one that is universally recognized in your field. It's vital, however, that you accept responsibility for taking material from a specific source. Do not deliberately rewrite something in your own words to avoid being accused of outright plagiarism. The fairest, most responsible solution is to acknowledge the borrowing with a footnote or signal statement like this:

> Roe, Soulis, and Handa emphasize that the fundamental distinction between analysis and synthesis depends on the distinction between separating a system into sections and unifying it into a whole [1].

Here you give the authors full credit as the source of the comparison, and you receive full credit for concisely summarizing the key distinctions. Writing experts say that the only way to write a summary (see Chapter 3) is to distance yourself from the original so that you don't unintentionally echo its phrasing. Yet, even when you represent an excerpt entirely in your own words, it is still plagiarism if you do not include a reference to the source. Consider the following paragraph from *Reality is Broken*:

> When we're playing a good game—when we're tackling unnecessary obstacles—we are actively motivating ourselves towards the positive end of the emotional spectrum. We are intensely engaged, and this puts us in precisely the right frame of mind and physical condition to generate all kinds of positive emotions and experiences. All of the neurological and physiological systems that underlie happiness—our attention systems, our reward systems, our motivation systems, our emotion and memory centers—are fully activated by gameplay.
>
> This extreme emotional activation is the primary reason why today's most successful computer and video games are so addictive and mood-boosting. When we're in a concentrated state of optimistic engagement it suddenly becomes biologically more possible for us to think positive thoughts, to make social connections, and to build personal strengths. We are actively conditioning our minds and bodies to be happier.

Now compare it with the following summary:

> There is a good reason why people become obsessed with gaming: physically and emotionally, they are completely engaged in a challenging and satisfying activity that makes them happier, stronger, and more positive.

This summary does not copy the language or point of view of the original, but the student writer is still presenting ideas that are not his or her own. Without a reference to the source in a note including a link to bibliographical information, such writing is nothing but plagiarism. To avoid it, add a signal phrase (author's name and credentials) to introduce the source fairly and then provide your summary followed by the citation.

> According to Jane McGonigal, well-known creator of alternate reality games, there is a good reason why people become obsessed with gaming: physically and emotionally, they are completely engaged in a challenging and satisfying activity that makes them happier, stronger, and more positive [2].

Plagiarism sometimes occurs by accident, but more often it is the result of unfair, unethical borrowing. With so much material available on the Internet and in libraries, dishonest people think no one will suspect that their work is not their own. They are usually mistaken. The writer who borrows indiscriminately is usually caught because the style of some sections is clearly inconsistent with the rest of the writing—a sure indicator that there is more than one writer at work. Experienced readers instantly recognize such discrepancies, to the detriment of the devious copier. To avoid any such suspicion, always do the following:

- Acknowledge any and all sources in notes and/or a bibliography.
- Use quotation marks around all direct quotations of words or phrases (see pp. 229–231). If the words or phrases are standard terminology, keywords, or nomenclature, you won't need quotation marks to include them in your discussion even though you do identify the original source in a note.
- Acknowledge in notes all figures and statistics cited, even though you do not put them in quotation marks.

If you are using anyone else's ideas to support your own, acknowledge that fact. Whether or not you name the author directly in the text ("As Porter confirms, . . ."), be sure to include the reference number and cite the source fully in your list of references. Never be afraid that your work will seem weaker if you acknowledge the ideas of others. On the contrary, it will be all the more convincing: serious academic treatises are almost always built on the work of preceding scholars, with credit duly given to the earlier work.

For students in a hurry, it is tempting to copy material from electronic sources on the assumption that it is in the public domain. Even though websites are instantly accessible, the material is the property of the individual or organization that published it, and it is protected just like printed material. Remember, then, that you must properly acknowledge the information you find on any website just as you would for anything from a book or journal. With search engines that easily detect plagiarism, you must never put yourself in a position which leaves you open to charges of wrongdoing.

Whether at school or on the job, you often collaborate on projects or assignments. Always list every contributor's name on the title page and give credit for special help in acknowledgements at the front of the paper. But bear in mind that when an academic situation calls for independent work, it is considered plagiarism to copy another person's assignment and present it under your own name, even if you worked on it together. While consultation is permitted, sometimes encouraged, duplication is never allowed. Given the penalties, letting someone copy your work is asking for trouble too. Collaborating with classmates is acceptable and may even be encouraged, but be sure to produce your own independent write-ups.

In any situation, your careful selection and documentation of material confirms your ability to take advantage of, while fully acknowledging, someone else's research. Above all, remember that anything you put in writing is there to convince your reader—especially someone who is going to give you a grade or a promotion—that you are methodical, precise, and professional.

Chapter Checklist

- ☐ Develop your own system to classify and file your notes for easy access and review.
- ☐ Become familiar with and take advantage of all the research and reference services provided by your school library.
- ☐ Create an account with the web-based citation manager provided by your school (RefWorks or other) to organize all your sources for projects.
- ☐ Signal and cite all sources accurately to maintain your academic integrity.

Notes

1. P.H. Roe, G.N. Soulis, and V.K. Handa, *The Discipline of Design*. Boston: Allyn and Bacon, 1972, p. 54.
2. J. McGonigal, *Reality is Broken: Why Games Make Us Better and How They Can Change the World*. London: Random House, 2012, p. 28.

Writing Summaries and Abstracts

Chapter Objectives

- Summarizing your own work and the work of others
- Creating an annotated bibliography
- Editing your summaries for brevity and clarity

Introduction

Among the writing skills invaluable to an engineer is the ability to strip something to its essence—to reduce something quite complicated to a few paragraphs, or even to a few words. You develop this skill through all the practice you get taking notes. You can also develop it deliberately through the exercise of learning to write an effective summary.

Some professors reminisce about learning the art of précis writing at university. (Writing a précis is the exercise of reducing a passage to exactly one-third of its original length.) This rather unnatural assignment helped teach people to make every word count, and many excellent writers attribute their talents to working on the précis. Today we teach students how to prepare a summary, and its function is not purely academic. From the abstract of a scientific paper to the executive summary prepared for the non-technical reader of a proposal or report, all summaries are designed to save time for the reader. A good summary lets a busy decision maker know immediately whether a document is worth reading. If you are the writer responsible for a major project, it makes sense to master the technique.

As a writer with important information to convey, you have good reason to make your summaries concise, accurate, and well-focused. Following the suggestions in this chapter will make the task of writing succinct, informative summaries easier.

Definition

Every summary is compact. Its purpose is to present, in the fewest, most precise words, the essence of a piece of writing. In an academic paper, the summary is called an *abstract*, and you write it for inclusion in a database or similar index. Its key feature is its length: almost all abstracts are expected to be shorter than 250 words (or, at most, one page of double-spaced text). You already depend on abstracts to help you determine which papers, articles, or dissertations support your research. In learning to compose an abstract for an academic project or a summary for your own engineering proposals or reports, you become better prepared to join academic and professional communities.

Summarizing Your Work

A summary is not the same as the introduction to a paper. It is a composition of one or two paragraphs that can be read and made sense of independently. The following rules are key to writing an effective summary:

1. **Prepare the summary or abstract once you have completed and edited your document.** Not only will you be working with a polished product, but you can be confident that the content will not change. Position the final version at the front of your report or paper, before your table of contents.
2. **Identify three to four keywords** (the field, your subject area, your special focus). Work them into the text of your summary to make them easy for a search engine to find.
3. **Remember to consider the needs of the reader.** Most readers will have the technical expertise to understand your subject, but there may be other readers who will not. If you know your audience, you are able to choose language that is suitable. Non-technical readers depend on the summary to answer two questions: *What is this?* and *What am I supposed to do with it?* Some readers read only the summary, so it's especially important that it be simple yet precise.
4. **Keep to the specified length.** Depending on the project, your summary or abstract may be strictly limited to a specific number of words, in which case you must meet those requirements exactly or risk having your work rejected. It makes sense to check requirements carefully before you begin. If you have not been given a word limit, match your

summary to the context. Most summaries should not be longer than a page of double-spaced text (maximum 250 words), but an executive summary for a major project may well extend to a second page.

Some professors assign extended summaries as part of a major project. Here's a rule of thumb for preparing these: using section titles as headings, produce a brief paragraph or two for each one. Include graphs or charts only as long as you can keep to the page limits for the assignment.

5. **Stress important findings.** Highlight specific discoveries or other new material you present in the complete text of your report or paper. If you use your summary to make your reader aware of these innovations, he or she will have good reason to read beyond the summary.

6. **Avoid generic statements about content.** It's tempting to develop the summary as a roadmap for the paper, a sort of table of contents in prose. Do not, however, waste the reader's time by writing something like "Findings are discussed extensively, and recommendations for further research are included." Readers already expect a paper to present findings and recommendations. They want to be guided to the specifics, not be told the obvious. Unless you provide them with a summary of the details, they won't see the need to read further.

   For a quick example, take a look at the summary that appears on the back cover of this manual. You'll see both an overview of the book's purpose and a bullet list of specific topics. If the context calls for more formality than is offered by a list of bullet points, reproduce these in one or two sentences. Other examples of summaries are the introductory statements of learning objectives at the beginning of each chapter or the single bolded sentences in this list of guidelines.

7. **Take advantage of conventional structures to make the writing easier.** Though your table of contents should not be the focus of your summary, it will serve as the best first outline for it, since it points you to the subjects for which you must account. One of the following models and sets of questions should help you stay on track.

- **the chronological summary**

PAST	Where did it begin? How was it done?
PRESENT	What is the current situation?
FUTURE	What lies ahead?

- **the experiment summary**

  PURPOSE   What is the hypothesis?

  METHOD   In what order were operations performed?

  RESULTS   What happened?

  CONCLUSIONS   What does it all mean?

- **the problem-statement summary**

  CONTEXT   What is the background of the problem?

  DEFINITION   What are the complications?

  OPTIONS   What are the alternatives?

  RECOMMENDATIONS   What action should be taken?

- **the proposal summary**

  CONTEXT   What is the background of the situation?

  CONSIDERATIONS   What basic requirements/specifications
                      are being addressed?

  NEEDS   What time/money/effort is involved?

  IMPLEMENTATION   What are the next steps?

Choose the formula that's most appropriate for your context. If your summary answers each of the questions in a sentence or two, it will be well focused. More importantly, it will address the reader's needs exactly.

Here's a one-paragraph summary of a 17-page project report. Organized in chronological order, it satisfies the reader's needs by answering three questions: *What? So what? Now what?*

In November 2014, the XYZ Film Lab in Toronto commissioned the design of a machine for chipping thin hard plastic for recycling. The goal was to conceptualize an alternative to the equipment currently used. The team developed a 2-cubic-metre prototype with a cylindrical blade like that of a rotary lawn mower. Stabilized on a drafting table, the prototype handled up to 10 sheets of plastic simultaneously, requiring sharpening only after 28 hours of operation. With its efficient size, safety features, and low power requirements, the new plastic chipper outperformed the existing equipment. Proposed research will lead to improvements in blade life, dust control, and noise reduction. (106 words)

8. **Do not include citations or references to other works.** The place for references to the work done by others is in the text of your document and in your bibliography, not in the abstract.

Summarizing the Work of Others

Sometimes you will be required to produce a summary of someone else's work, either as an academic exercise or on the job as a backgrounder for a busy colleague. This summary isn't like a review: it is not intended to evaluate, comment, or criticize.

Your summary of someone else's work simply records, as accurately as possible and in as few words as possible, your understanding of what the author has written. Whether you like what you have read is not the issue. Your job is to get to the heart of things—to separate what is important from what is not.

1. **Determine the author's purpose.** Every author writes for a reason: to cast some new light on a subject, to propose a theory, or to bring together the existing knowledge in a field. Whatever the purpose, you have to discover it if you want to understand what guided the author's selection and arrangement of material. The best way to discover the author's intention is to check the preface, the introduction, and—if there is one—the author's own summary or abstract. If such a comprehensive summary exists, avoid borrowing from it in a course assignment. If you're on the job, don't reinvent the wheel: present that summary to the person who commissioned it and acknowledge the source. A glance at titles and subtitles, headings and subheadings shows you what the author considers most important and what kind of evidence he or she presents. The details are much more understandable once you know the direction of the discussion.

2. **Read carefully and take notes.** A thorough reading will be the basis of your note-taking. Since you have already determined the relative importance that the author gives to various ideas, you can be selective and avoid getting bogged down in less important details. Just be sure that you don't neglect any crucial passages or controversial claims.

When taking notes, try to condense the ideas. Don't take them down word for word, and don't simply paraphrase them. You will have a much firmer grasp of the material if you resist the temptation to quote. Force yourself to summarize. This approach also helps you be concise. Remember: you want to be brief as well as clear. Condensing the material as you take notes ensures that your report is a true summary, not a string of quotations or paraphrases.

3. **Follow the same order of presentation as the original.** It's usually safer to follow the author's lead. That way your summary is a clear indication of what's in the original.

4. **Discriminate between primary and secondary ideas.** Give the same relative emphasis to each area that the author does. Don't just list heading or chapter titles or reiterate conclusions.

5. **Include the key evidence supporting the author's arguments.** Include supporting details. Without them, your reader will have no way of assessing the strength of the author's conclusions.

6. **Use bullet points when possible.** They help your reader see relationships.

Writing an Annotated Bibliography

Some research assignments call for an annotated bibliography, which is a list of sources accompanied by a brief description of each item on the list. The Institute of Electrical and Electronics Engineers (IEEE) provides the following example:

[B10] Henry, S. and Selig, C., "Predicting source-code complexity at the design stage," *IEEE Software*, vol. 7, no. 2, pp. 36–44, Mar. 1990.

> *This paper states that the use of design metrics allows for determination of the quality of source code by evaluating design specifications before coding, causing a shortened development life cycle.* [1]

Note that this example identifies not only the type of publication (*paper*) and contents (*design metrics*) but also the timing (*evaluating design specifications before coding*) and implications (*a shortened development life cycle*)—all in one sentence.

If you are asked to provide an annotated bibliography, either as part of a project or as an independent exercise, here's what to do. Begin by alphabetically arranging your entries for the bibliography and providing full details according to the referencing system you are following (see Chapter 9 for options).

Then add your own summary (one to three sentences) of the contents of each source, including information of value to your reader. Essentially, you answer the same questions you ask when preparing an executive summary: *What is this?* and *What can we do with it?*

Use your own words and phrasing in answering this question. It is tempting to reproduce the document's own summary or abstract or the "about us" introduction that appears on a home page, but expectations of academic integrity and professionalism (see p. 26) require you to create your own sentences. Here is an example of an annotation in IEEE style:

Canadian Wind Energy Association Website [online], <http://canwea.ca> (current Aug. 15, 2014).

> With federal support, this website provides up-to-the-minute information on Canadian projects powered by wind energy. In addition to promoting conferences and membership, the website highlights the potential of wind power as an economical energy alternative. It includes a sophisticated cost/savings calculator that matches wind conditions to turbine requirements allowing home and business owners to complete their own cost-benefit analysis. [60 words]

Editing for Economy and Precision

Whether you are writing a summary of your work or someone else's, it is essential to leave time for thorough editing. These two final guidelines are vital:

1. **Reread and revise your summary to make sure it's coherent.** Summaries can often seem choppy or disconnected because so much of the original is missing. Use linking words and phrases (see p. 90) to help create flow and give the writing a sense of logical development. Careful paragraph division will also help to frame the various sections of the summary.
2. **Revise to make every word count.** You may find that you have to edit your work a number of times to eliminate unnecessary words and get your summary down to the required length. Be ruthless about eliminating deadwood (see p. 167), and avoid passive sentences. If you find it impossible to reduce the length any further, try starting over rather than picking at words. It may be easier to generate another set of answers to the model questions than to be constrained by a bulky first draft.

Chapter Checklist

- ☐ Identify the audience for your summary.
- ☐ Read the original carefully, noting what your audience needs to know about the subject and its context.
- ☐ Prepare a single sentence summary of the contents. Then, flesh out your summary by answering the following three basic questions: What? So what? Now what?
- ☐ Keep to the assigned length.
- ☐ Take the time to edit thoroughly.

Note

1. "Annotated bibliography," (2005, Jan.). *IEEE Standards Style Manual*, (rev. ed.) [Online]. Available: http://grouper.ieee.org/groups/1057/2000Style.pdf (current Aug. 22, 2014).

Writing a Lab Report

4

> **Chapter Objectives**
> - Understanding the purpose of a lab report
> - Organizing the sections of a lab report
> - Writing your results

Introduction

A major part of the work you do as a student involves learning the fundamentals of research and analysis. This is one reason that you are expected to attend labs and prepare reports to turn in for credit. When you graduate to an engineering career, knowing how to write and interpret technical reports will be essential to the work you do. You have plenty of opportunity for practice at college or university.

All lab assignments demand accuracy and objectivity in recording and presenting what has been done. Engineers are interested in exact information and the orderly presentation of supporting evidence. Although you may wish to make a case for a particular hypothesis, professionalism obliges you to separate the facts you are reporting from your own speculations about them. Never allow preconceived opinions or expectations to interfere with the way you collect or present your data; if you do, you risk distorting your results. Conduct your experiment as objectively as possible, and present the results so that anyone reading your report or attempting to duplicate your procedures will likely reach the same conclusions you did.

Purpose

As a student in a lab, you do work and then prepare reports to demonstrate that you can apply a theory or know how to test a certain hypothesis. The person who marks your work already knows the methodology and the nomenclature. But while there is no need for you to explain your terms, you must still make it clear that you understand what they mean. Your reader will be on the look-out for any weaknesses in method or analysis and any omissions of important data. Usually you are expected to give details of your calculations, but even when you have been asked to provide only the results of these calculations, you should still identify and justify irregularities that might affect their accuracy.

Format

Organize your report into separate sections, each with a heading. By convention, most reports follow a standard order:

1. *Title page*
2. *Abstract* (or *Summary*)
3. *Contents* (for extended reports)
4. *Introduction* (or *Purpose* or *Objectives*)
5. *Materials* (or *Equipment*)
6. *Method* (or *Procedure*)
7. *Results* (or *Observations*)
8. *Discussion* (or *Analysis*)
9. *Conclusions*
10. *References*
11. *Appendices*

The order of these sections is always the same, although some sections may be combined or given slightly different names, depending on how much information there is in each one. Note that this order of presentation doesn't necessarily reflect the order in which you complete the lab work or, indeed, the order in which you do your write-up. From the notes you have taken during the lab (see p. 19), prepare a draft of your report using the section headings above as a template. If your instructor expects lab work to be recorded in a bound notebook turned in for grading, your submission may not include all the formal sections demanded of a technical report, but it still calls for answers to specific questions and requires data to be recorded accurately and analyzed in the context of the course. Different departments have different expectations, and it makes sense to follow a preferred template. The following presents an overview of each section of a formal lab report.

Title page

The first page of the report should include your name, the title of the experiment, the date it was performed, and the date you turn in the report. For practical purposes, it should also include the name of your course and instructor. If you are using a bound lab book, course information will occupy the top half of the first page in the book, followed by a few blank pages on which you will record titles and dates of labs (and their page numbers) as you complete these.

Titles should be brief (10 or 12 words) but informative, and you should make sure they clearly describe the topic and scope of the experiment. Avoid meaningless phrases, such as "A study of . . ." or "Observations on . . .". Simply state what it is you are studying, such as "Flow Rate Tests for 6V Pumps."

Abstract

The *Abstract* appears alone on the page following the title page (or on the lines following the title in a lab book). An abstract, as explained in Chapter 3, is a brief but comprehensive standalone summary of your report. Anyone should be able to read it and know exactly what the experiment was about, as well as what the results were and how you interpreted them. Remember that economy and precision count: for a simple experiment, keep the summary shorter than 75 words; for a complex one, make the maximum length 150 words. Of course, to achieve these limits you must avoid vague or wordy phrases, such as "The reason for conducting the experiments in this study of X was to examine . . .," when "The study of X examined . . ." is only one third as long.

Contents

If the report is lengthy, extending over several pages and including attachments as appendices, provide a *Contents* page that lists section numbers, titles, and page numbers. In extended reports, you also include a *List of Figures* and a *List of Tables* on separate pages following the *Contents* page. If you are writing in a bound notebook, it is especially important to include a table of contents and number all your pages, so that the person marking your lab can easily find where to start.

Introduction

In the *Introduction*, give a detailed statement of purpose for the experiment you have undertaken. Describe the problem you are studying, your reasons for studying it, and your research strategy for obtaining data. If, as is often the case, your purpose is to test a hypothesis, you should state both the nature of the problem and your expectations of the findings. The introduction should also include the theory underlying the experiment and any pertinent background data or equations. Although you may refer to outside sources—especially if you want to add legitimacy to an experiment you've developed yourself—you should still put the emphasis on your own work and its context. If you are completing the experiment as part of a course, your introduction will be a statement of purpose that's only a sentence or two long.

Materials

The *Materials* section presents a description of the materials and equipment you used and provides some explanation of how you set up the experiment. Often you can include this section as part of your discussion of procedures in the *Method* section. If you did any manipulating of the equipment at different points in the experiment, give a full list of equipment here, and describe each separate arrangement.

A simple diagram or two—produced by computer or by hand—will help the reader visualize your arrangement of the equipment. If a diagram is too complicated to fit a regular page, include it as an appropriately labelled attachment at the back of the report and direct the reader to it with a reference.

Even if the apparatus or materials you are using are standard, commercially available items, note the name of the manufacturer, the model number (if applicable), and the name of the source or supplier: for example, "spectroscopic-grade carbon tetrachloride (99% pure) supplied by BDH Chemicals in St. John's."

Method

The *Method* section is a step-by-step description of how you carried out the experiment, with procedures presented in the order you performed them. If your experiment consisted of a number of tests, begin this section with a summary identifying how many tests you ran. This way, your reader is prepared for the numbering of your series. When you describe the tests later in the report, use the same numbering system to prevent confusion.

Write this part of the report with enough specificity to permit anyone to duplicate the experiment in all its details. If you are following instructions in a lab manual, include page references. Summarize the instructions in your own words rather than copying directly. Verify conventions for acknowledging such sources with your lab instructor. Often a standard bibliographic reference is all that's required (see Chapter 9).

Although you should be concise in your description of the experimental method, make sure that you don't omit essential details. If you heated the contents of a test tube, for example, be sure to report at what temperature you heated them and for how long. If you performed a chromatography or other process at a faster or slower rate than usual, it's important to indicate the rate. Readers need to know exactly what controls to apply if they try to perform the experiment themselves.

When reporting the results of experiments, it is standard practice to use the past tense. However, scientists regularly debate whether to use the active or passive voice (e.g., "*I tested* the sample" versus "The sample *was tested*"). Traditionally, the passive voice was preferred because its detachment seemed appropriate for scientific contexts (see pp. 169–170). More recently, writers have tended to use the active voice because it is less likely to produce convoluted sentences. Ask an instructor about your department's preferences, but use your own judgment about what sounds best. Your goal, whichever voice you use, is to be clear, concise, and objective. When people write well, readers don't take exception to their phrasing.

Results

The *Results* section is the section of most interest and value to experts, and they depend on its accuracy. It usually presents a blend of data and description. It may also contain statistical calculations.

Find out from your instructor whether you are expected to give the details of your calculations or only the results of those calculations. (You can always include extensive calculations in an appendix.) Take time to double-check your numbers. You should also make sure where possible that the calculated values you report include any *uncertainty*. For example, you might report that the calculated volume of a hollow sphere is 23.45 ± 0.05 cc (where ± 0.05 is the uncertainty in the volume measurement). When reporting any calculations or measurements, check to see if you need to include the standard deviation, the standard error of the mean, or the coefficient of variation.

The format of your *Results* section depends on the type of experiment you have performed. Generally it begins with your main findings and then deals with secondary ones. Whenever practical, summarize your results in a graph or table. (Graphs and charts generally have greater visual impact than a table, but the table is more useful if you have made several measurements.) Whatever graphics you use, label them clearly. Make sure that you refer to and explain each figure or table. You must not expect such material to speak for itself.

Pay attention to the following guidelines when you are creating charts or graphs for your lab report (see Chapter 10 for additional information):

- Use a scale that will allow you to distribute your data points as widely as possible on the page.
- Use large and distinctive symbols, with different symbols for each line on the graph.
- Put error bars (±) on data points, where necessary.
- Label the axes clearly, and always include the units of measurement used so that the reader knows exactly what you have plotted on the graph. Always include a legend to identify the units or to explain what different symbols represent.
- Label the graph (for example, *Fig. 1*) so that you can refer to it by name and number in your report. If you have more than a couple of these, remember to include a *List of Figures* after the table of contents at the beginning of your report.

Discussion

The *Discussion* or *Analysis* section of the lab report allows you the greatest input; it is here that you interpret the test results and comment on their significance. You want to show how the test produced its outcome—whether expected or unexpected—and to discuss elements that influenced the results. In determining what details to include in the *Discussion* section, you might address the following questions:

- Do the results reflect the goal of the experiment?
- Do these results agree with previous findings, as reported in the literature on the subject? If not, how do you account for the discrepancy between your data and those accepted or obtained by other students and researchers?
- What may have gone wrong during your experiment, and why? Can you propose a source of error?
- Could the results have another explanation?
- Did the procedures you used make sense in light of what you hoped to accomplish? Does your experience suggest a better approach for next time?

For a good discussion, remember to think critically about how your own work relates to previous work that you have read about or done yourself. Remember to acknowledge primary and secondary sources.

Conclusions

The *Conclusions* section briefly lists the conclusions that may be drawn from the experiment. You don't necessarily need a separate section for these—they may also appear in a short paragraph at the end of the *Discussion* section.

References

List references to any outside sources named in your experiment, including the course textbook, on a separate page before the *Appendices*. For details on documentation format, see Chapter 9.

Appendices

If you have extensive detailed results, particularly computer-generated ones that fill more than a page or two, include these at the end of your report. If you have more than one section, number them *Appendix 1*, *Appendix 2*, and so on. Make sure each has a title that identifies its contents (for example, *Appendix 2: MATHCAD Worksheet*), and list them, including the titles, in your *Contents*. In a lab book, you will simply paste these data or calculations in where they fit naturally (as suggested on p. 20).

Chapter Checklist

- ☐ Confirm with your professor or supervisor the format and order expected of a lab report in your discipline.
- ☐ Make sure you include all relevant sections so that your report addresses the following: statement of purpose, methodology, observations, and conclusions.
- ☐ Proofread and edit your lab report carefully before submitting it.

Writing Proposals and Project Reports

Introduction

Much professional engineering work is straight problem solving. Whether you are working independently or collaborating as a member of a team, your professional life will often require you to prepare full-scale reports on work that has been done—or on work that you want to be engaged to do.

As a student, you have opportunities to develop your writing and research skills in preparing reports on major research or design projects on campus. If you are involved in work/study projects as a co-op, internship, or service-learning student you will soon gain experience in writing at work. The guidelines presented in this chapter offer general strategies and techniques for writing effective proposals and reports both in school and on the job.

Before You Begin

Remember that people with demands on their time want to know as quickly as possible the core message of anything they are reading. They also expect to be able to trust its accuracy. There are three basic principles to keep in mind at all times:

1. **Put the most important information up front.** No matter what you are writing, always include a clearly identifiable summary. But a business context also calls for a letter of transmittal, sometimes referred to as a cover letter. Adopting different phrasing from the summary

(to make it clear that you've taken the time to do so), the letter of transmittal establishes the context for your document and highlights the action you are recommending.

2. **Be concise.** Everything you write should say as much as possible in as little space as possible.

3. **Be objective.** Readers must be confident that the information you are providing is professionally justified and free of any bias, ulterior motive, or hidden agenda.

Planning

What you write varies according to where you are in the problem-solving process. The following kinds of reports suit different stages of this process:

Problem statement

A problem statement can be a fairly extensive discussion to introduce and document a problem you have identified. An academic context can require you to examine four separate considerations: context, definition, constraints, and criteria (see p. 56). In your engineering courses, a problem statement may be an independently evaluated assignment. More likely, however, it will find a place at the front of your final project report.

Proposal

Once you have a firm definition of the problem, you need to identify solutions, evaluate alternatives, and determine the process it makes sense to follow. Your goal is to get the go-ahead to embark on the project. Thus you have to develop an extensive proposal that details, section by section, the process you will follow, the equipment and staff you will require, the timelines and costs for the project, and, most importantly, the results you expect. In an academic context, a research proposal generally establishes the theoretical context for the work and includes a tentative outline for the final write-up.

In the professional world, the emphasis is on results. A proposal is accepted only when it is deemed to meet the needs of the organization better than any other proposal. It is vital, therefore, that your proposal anticipates and addresses all issues of concern.

There are two separate contexts in which you might prepare a proposal in a business setting. If you have been invited to present a proposal, look to the organization of the Request for Proposal (RFP) to spell out the detailed content and arrangement of material you are expected to supply. If, on the other hand, you are developing a proposal independently, you have to make an even stronger case. A company that sends out an RFP already understands that there is work to be done. But when you are the one making contact, with an unsolicited proposal, your problem statement and solutions must make it obvious that the work is necessary. In any case, if you want your proposal approved your writing must be clear and convincing—and your format must be professional.

Progress report

Once a proposal has been accepted and a project begins, you are expected to keep people informed of your progress. These regular updates may be informal (for example, you will send them by e-mail), but they are important enough to spend time on writing well. Use them to emphasize not just what work has been completed, but also where the project stands with respect to the overall approved plan. If there are any deviations from the original, account for these immediately. If schedules or budgets have to be revised, report the situation thoroughly and responsibly. Even when everything is going according to plan, it is your responsibility to keep everyone informed. There's no need for an extensive discussion in this situation—a quick message to prevent your reader from worrying why he or she hasn't heard from you for a while will do. Whether your progress report is extensive or brief, therefore, depends on its message.

Final project report

Once a project is complete and all the assigned work is done, prepare a final comprehensive report as a permanent record. Make sure you allow time in your original project schedule for the production, editing, and distribution of this final report.

Defining Your Task: The Four *Rs*

Taking time to organize your thoughts and to devise strategies is well worth the effort. Wherever you are in a project, define your task precisely by asking questions about four aspects: *reason, reader, restrictions*, and *research*.

What is the reason?

Why are you writing? What goal are you trying to achieve? Broadly speaking, you write for one of two basic purposes: to provide information or to recommend some course of action. Many informational reports (like progress reports, production reports, and monthly sales reports) are designed to pass along facts and data as they accumulate. So that they don't strike the reader as bureaucratic busywork, make sure they focus on exceptional rather than routine matters. On the other hand, reports written to make specific recommendations—to help someone make a decision or to propose an alternative—usually receive close attention. Examples are a feasibility study, a proposal for a new product design, or a suggestion for making a process more efficient. This kind of writing provides the most opportunity for you to show your analytic ability and creativity—clear evidence of professional competence.

Determining your reason for writing allows you to establish both a purpose and an expected outcome. If an important decision rests on the document, you will have to consider exactly what information is needed to make that decision and precisely how you will support any recommendations.

Who is the reader?

Although several people may read your report, you usually have only one primary reader. Knowing who this reader is helps you organize and present your material so that it's likely to be well received. If the project is assigned in a course, the person grading your work will likely be your professor or supervisor, but the assignment itself might specify a business audience or other professional context. Make sure you define the intended reader carefully. If you are doing graduate work, there may also be a committee or an external reader. In professional contexts, the reader might be a project manager, a department head, a CEO, or a board of directors.

You can always expect your primary reader to be professional, demanding, highly motivated, and concerned about the bottom line. The more you can identify his or her priorities, the better you will be able to meet them. When there are secondary readers, count on your primary reader for advice on writing for them. Among the details you should consider are these:

- **What is your relationship to the reader?** Is the reader your boss or a colleague? How has he or she reacted to past communications with you? If you are writing for someone in a position higher than yours, your tone and approach should be more formal than they would be for an associate you often work with. Always remember to keep things readable, though (see Chapter 13).

- **Has the reader asked for the report?** If you are writing a proposal in response to a request, you may not need to fill in much detail about the purpose. If what you are writing is unsolicited, you must establish a compelling context for your recommendations.
- **What is the reader's area of expertise or responsibility?** You need to go into more detail about an area that is your reader's specialty than you do for one in which he or she is less interested or involved. In business, top management wants an overview, whereas a specialist requires all the particulars. Remember that your report might be going to several different kinds of readers—to a plant supervisor as well as to your boss, for instance. If so, prepare the complete analysis for the supervisor with an executive summary designed to be read by management.
- **How is the reader likely to respond?** Consider the reader's situation and the expectations and concerns that he or she is likely to have. Are you delivering good news or bad news? (It's usually good news if it can save time, money, or effort; it's bad news if it means more time, money, or effort than expected.) If you can anticipate objections or concerns and address them in your writing, your solutions will be much more convincing.
- **How might the reader benefit from the report?** Your suggestions will always be more persuasive if you can point out their advantages (good news) for the reader. The benefit could be significant: giving the reader a competitive edge, for example, or saving the business money. Even if the benefit is a more general one, such as improving the ability to anticipate and manage future problems, you should point it out.

What are the restrictions?

From the outset, consider the practical restrictions on your own writing. For example, how much time do you have? How much help is available for preparing the final report and for producing illustrations, blueprints, or prototypes?

Other restrictions will apply to the subject of the report. If you are asked to choose your own project, be sure to narrow it to a manageable size. On the job, however, you will likely be given a predetermined project, one on which you may be the chief consultant, but still one with clear limitations. Since it's always better to do a thorough job on a narrow subject than a superficial job on a broad one, invest the time to limit your topic to manageable proportions.

What research *is required?*

In deciding what information to gather, weigh the time and money required to do the research against its usefulness to the report. In other words, determine what is essential.

It is especially useful to identify how much your reader already knows; in many cases your reader will not need any background information at all. If you must provide some, remember that too much detail may draw attention away from more important matters. One solution to this problem is to put non-essential details in appendices at the end of the report.

Once you have decided on the information you need, ask yourself what cross-checking you have to do. Are your sources reliable? What facts or figures do you need to verify? Use your engineering expertise to determine the degree of accuracy or precision required for any figures you supply. Indicating the margin of error will show the reader that you are thorough and objective.

If you are on the job, the facts and figures you need can usually be obtained either by questioning people or by researching company documents. If you find that an earlier report covers some of the same material you are working on, it's important to refer to it, updating facts and figures as necessary. If more extensive research is required, consult government documents, company reports, or academic publications. Most companies' websites give easy access to annual reports, corporate information, product announcements, and financial information. On the job, just as at school, be sure to acknowledge all your sources and give proper references (see Chapter 9 for instructions on documenting your sources correctly).

If you obtain your own data—through experiments, tests, or surveys, for example—make sure that any results are based on an appropriate sample. If you aren't familiar with proper sampling methods yourself, consult someone who is. Nothing weakens credibility more than providing outdated, unreliable, or invalid statistical information.

Developing Your Discussion: Three Common Patterns

The way you develop ideas depends on your purpose and the order in which you decide to present them. The origins of a problem, for example, are often presented in chronological order. When you want to convince a reader to accept your recommendations, however, most of your content will follow patterns for *defining*, *listing*, or *comparing*. Here are some suggestions for taking advantage of these three patterns.

Defining

Sometimes you want to explain the meaning of a term that is complicated, controversial, or simply important to your field of study: for example, *bias* in measurements, *ergonomics* in design, *dendrites* in geology, *fidelity* in electronics. Often, all it takes is a single sentence, or even just a quick definition in parentheses. But sometimes a definition becomes a separate assignment or a question on an exam. It's worth recognizing the requirements.

If you decide a definition is necessary for your context, perhaps because you are preparing a problem statement, begin with a general statement to introduce the term. Then make your definition more exact: it should be broad enough to include all the things that belong in the category but narrow enough to exclude things that don't belong. A good definition builds a kind of verbal fence around a word, herding together all the members of the class and cutting off all outsiders.

For anything beyond a bare definition, include illustrations or examples. Depending on the nature of your discussion, these could vary in length from one or two sentences to several paragraphs. If you are defining *cybernetics*, for instance, you may want to discuss at some length the various so-called sciences of complexity, including neural networks and artificial intelligence.

In an extended definition, it's also useful to point out the differences between the term you're defining and any others that may be connected or confused with it. For instance, if you are defining *deciles* in statistics, you will want to distinguish them from *quartiles*; if you are defining *precision*, you should distinguish it from *accuracy*; if you are defining *trademarks*, you may want to distinguish them from *patents* or *logos*.

Listing

One way to organize material persuasively is to follow the current trend of *prioritizing* information, arranging things in their order of importance. At the heart of this structure is the bullet list, which can introduce any lineup of details: causes, effects, criteria, constraints, objectives, advantages, disadvantages, costs, findings, or recommendations. The attraction of this approach, which also governs what appears in the executive summary, is that readers recognize priorities right away. There's no time wasted.

Whenever you can simplify material by listing it, do so. A list, like a heading, is an aid to a quick understanding of any sequence of three items or more.

If you will be referring to the items in the list later, number each one; otherwise use a simple bullet (•). Whenever you list items, make sure all of your points are consistent and parallel. For example, if you phrase one point as a complete sentence, make all of your points complete sentences. Or if some of your points begin with verbs, make all of them begin with verbs. If your list is introduced with an incomplete sentence, make sure each point in the list properly completes the sentence. (See pp. 199–201 for additional information.)

The principle of "important things first" also applies to the arrangement of points within each section. In the *Recommendations* section especially, it's a good idea to put your most important point first, followed by the remaining points in descending order of importance.

Throughout, keep your purpose and reader in mind. For example, if you are trying to persuade a hostile or skeptical audience, you may find it better to begin with the point, major or minor, that will get the most favourable reaction.

Comparing

When you are presenting information and recommending a choice, you should organize your material in a way that helps the reader understand the options. For example, suppose that you have been asked to recommend one of three LAN models. The comparisons will be clearer for your reader and an assessment easier to make if, instead of describing each model in turn, you use a single paragraph to define the criteria—processor speed, memory, connectivity, and price—and then assess all three models together with respect to each criterion. This kind of comparison is represented most logically and concisely in a table, which you supplement with a paragraph or two justifying your recommendation. In other words, after defining the criteria, you present the results of the comparison explaining the advantages of each choice and concluding with your recommendation.

Whichever way you choose to arrange the details, remember to be systematic and consistent. In addition, if you include a chart or table to illustrate the points of comparison, make sure you add a summary paragraph of explanation. Never expect visuals to speak for themselves.

Organizing the Parts

Normal proposals and reports include a number of conventional parts, whether the context is academic or professional. You determine which parts to include according to the extent of your project, its purpose, and the expectations of your audience. Most academic institutions and professional organizations have specific expectations of the written work you do for them, so it makes sense, right from the start, to confirm what these are. Most organizations have templates to help you create standard documents from memos to project reports. These make it wonderfully easy to structure and format work consistently. It would be silly not to take advantage.

A progress report may be nothing more than an e-mail message or memo with a line or two of copy. More formal documents, however, call for some or all of the following elements.

Front matter

The following material appears at the beginning of any formal document, from an extended project report to a doctoral dissertation. Use lower-case Roman numerals to indicate the page number at the bottom of the pages of front matter. You don't need to number the first page (which likely features the executive summary). Do not include the title page or the letter of transmittal in your numbering.

Letter of transmittal

The letter of transmittal (or cover letter) is a conventional letter or memo addressed to the person or group commissioning the proposal, project, or report. It is usually clipped to the front of the document, but it may also be placed directly after the title page and before the *Summary*.

This letter has four sections. You begin with a sentence or two establishing the context for the attached document; your next paragraph highlights the main contents. Reserve the third paragraph to acknowledge the contributions of people who helped you with the work. (If you are preparing such a letter to accompany a student project, you may also need to include a statement asserting that you are the sole author.) The final paragraph ends politely and conventionally, with a specific request for action, if called for. Throughout, it's appropriate to use personal pronouns like *I, you,* or *we*. Here is an illustration:

> In September, you asked the Engineering Students Association to investigate campus security. We are pleased to enclose our report.
>
> Our main recommendation is that the Student Federation, with the co-operation of the Student Volunteer Bureau, establish an evening escort service to safeguard students walking across isolated areas of the campus at night. This service would benefit not only part-time and extension

students going home from classes but also students who need to work late in the library or labs.

We are grateful for the support of your administrative assistant, Felicia Penofsky, whose suggestions were extremely helpful to us in conducting our investigation.

When you have had the time to consider our recommendations, we would welcome the chance to discuss them with you.

Title page

The front page features the title of the document, the name of the company or organization for whom the work was done, the names of the person or people who prepared the document, the academic affiliation, and the date of submission. If you're writing a course assignment, don't forget to include the necessary information, such as course name or number. It's always best to include your student ID number as well.

Abstract/Summary

Place your summary on a separate page directly after your title page. In formal academic papers this is called an *Abstract*, but you can use the designation *Summary* for most situations. In business contexts, call it the Executive Summary.

Keywords

If you are preparing an academic article for a journal, you will probably be asked to provide an alphabetical list of up to 10 terms that will enable someone doing a search to find your paper in a database. These are the same keywords that you use when you are conducting your own library searches. You may be asked for a similar list when you prepare a major project or assignment at college or university. Check with your course instructor or your library for examples of typical keywords in your field, or consult the IEEE Taxonomy available online. [1]

Table of contents

Readers look to the table of contents for immediate guidance to specific information. Formats may vary. To make yours as clear as possible, number each section and subsection of your document, and record these numbers together with their titles in the table. Indent subheadings, and place the page numbers for all sections and subsections in a corresponding column on the right-hand side of the page. Where possible, take advantage of auto-formatting options offered by your word processing application.

List of tables/List of figures

If your project contains tables or illustrations, or both, include a separate table of contents for each list. If there are not many of these, the two lists may share a page; otherwise, keep each to its own page.

List of symbols

If you have used symbols that are not common enough to be recognized or understood by your primary or secondary readers, list them on a separate page here.

Definitions/Nomenclature

Sometimes you will want to include a list of definitions of key terms. You can insert such a list here, or you can put it at the end of your work, in which case you call it a glossary.

Preface/Acknowledgements

In longer documents, and especially for projects that have received financial backing, it's appropriate to acknowledge any assistance in a few sentences or paragraphs at the beginning of the document. This acknowledgement goes in a section entitled *Preface* or *Acknowledgements*. Although sometimes you use the preface to establish a context for your investigation it does not fill the entire role of an introduction and cannot replace one.

Main Body

Introduction

This section may include a statement of purpose, a discussion of the background, context, or reason for the report, and/or an explanation of the method used to gather the information.

Problem statement

If you are writing a proposal or an investigative report, your reader might find it useful if you present a short summary of the problem, along with a discussion of the *constraints* (the limitations imposed on the solution) and the *criteria* (the features or characteristics required of the solution). Sometimes the constraints and criteria have been provided in project specifications or an RFP, in which case you might include them in an appendix instead of re-peating them here.

Discussion

By far the most extensive part of your proposal or report, this section contains the specifics of your investigation and is generally referred to as "the body." In it, you lay out detailed information, section by section, according to your pro-ject plan. Give each part a meaningful, yet brief, title: *Overview*, *Methodology*, *Site Visits*, *Design*, *Test Case*, *Results*, *Refinements*, and so on.

If you are preparing a problem-solving proposal, you include all the spe-cifics of your timelines and budgets in addition to your objectives and the procedures you expect to follow. Remember that a proposal is written before a project is undertaken; you look efficient if you account for as many contin-gencies as possible. On the other hand, a final project report is submitted after all the work has been completed, whether or not things took place according to plan. Part of your discussion in a final project report considers how well the results met the requirements set out in the original proposal. Details of failures as well as successes are expected. In fact, the discussion section often contains, near the end, a part entitled *Findings*, in which you discuss the de-tails of your results, both expected and unexpected.

Conclusions/Recommendations

Conclusions are the inferences you have drawn from the findings, and recom-mendations are suggestions about what actions to take. Depending on your reasons for writing, you may have conclusions or recommendations or both in this section, so the title you give varies accordingly. If you separate conclusions from recommendations, however, be sure to provide two separate sections.

Back Matter

After the main body, include one or more of the following sections, as called for.

References

The *References* section documents all material, published or unpublished, that you have cited in the report. You have several options for formatting this sec-tion, as outlined in Chapter 9. You are, however, expected to follow the conven-tions of your organization or academic institution. Check with your employer or instructor to find out if there is a particular model of documentation that you are expected to use.

Bibliography

If you want to recommend books, articles, and other materials that you have consulted but not cited or referred to in the report, the place to do so is here. Follow the same format you used for your references, but here you can also add a sentence or two summarizing the contribution of each work (see p. 36 for sample annotations).

Appendix/Appendices

An appendix contains any material that substantiates claims made in the body of your document but that you have not, for various reasons, included there. Appendices might include additional tables, questionnaires from a survey, summaries of raw data, parts of other reports that are pertinent to your findings, detailed spreadsheets, or any other information that would be an unnecessary distraction if it appeared in the main text. Keep separate appendices for each different type of material, and be sure to give each one a name and a place in your table of contents. More importantly, you must be sure to make reference somewhere in the text of your discussion to the material in the appendices. It's a mistake to include an appendix that has no clear connection to your project.

Finalizing Your Work

Include illustrations

Tables, charts, and other illustrations—especially those that analyze a quantity of data—are common and useful in any kind of report. Such visual aids help the reader grasp information that would take many words to explain. Use visual aids wherever you can to summarize, or pull together, information—but not simply to repeat what you have said in words (see Chapter 10 for guidelines on using effective visual aids).

When determining the best place for your tables and figures, begin by identifying the importance of the information you are trying to present. If a chart or table contains supplementary information, it belongs in an appendix. If the information is germane to the discussion, position the illustration soon after you first mention it. Number and label it appropriately. At the same time, be sure that you refer in your text both to the illustration and to the point it makes: for example, "As Fig. 2 shows, costs have decreased for each of the last five years."

Follow conventional formats

It is easy to format documents, so take advantage of what's available, as long as you know it's acceptable to your audience. The LaTeX documentation preparation system [2] and other document preparation applications simplify typesetting, formatting, referencing, and even indexing.

If you don't have access to these time-savers, follow recognized conventions for numbering and layout, which help your reader understand the relative importance of each of the sections in your document. Here are some suggestions for producing a report with a format that complements the professionalism of its content:

- Technical reports prefer the decimal system, which uses 1.0 and 2.0 to designate primary headings, 1.1 and 1.2 (after 1.0) for secondary headings, and 1.11 and 1.12 (after 1.1) as the third level. Other numbering systems include alphanumeric (A, B, C, 1, 2, 3, a, b, c) or Roman (I, II, III, 1, 2, 3, i, ii, iii). Whichever method you choose, make sure you use matching symbols for sections of equivalent importance.
- Leave a generous amount of room around the text to create an impression of space; wide margins and spaces between sections help break up an imposing mass of type, especially when it's single-spaced. (If you are preparing an assignment for a college or university course, ask if your instructor prefers double-spaced text; many academic readers have strict format requirements.)
- Make headings stand out by using bold or uppercase letters. But don't go overboard by using colour or fancy typefaces that only distract the reader from the essential content you are trying to present.
- Use different alignments and font sizes for different levels of headings. Systems for formatting headings vary, but the following is one convention that's widely used for technical papers:

SECTION HEADING

CENTRED AND IN UPPERCASE 14-POINT LETTERS

Secondary Heading at Left Margin

The heading is in 13-point bold type. The initial letter of the first and other important words is in uppercase, and the text begins on a new line.

Third-level heading. Punctuated as a sentence, the heading is in 12-point italics, and only the first letter of the first word is in uppercase. The text follows on the same line.

No matter what style you decide on, remember that the watchwords for headings with visual appeal are *clarity* and *consistency*.

Have your work professionally bound

In business, formal presentations call for professionally bound proposals or reports. Make sure you leave enough time to have this work done. Extended academic projects look better bound as well, but check with your instructor before you go to unnecessary expense. Some professors want hard copies of assignments turned in with only a simple staple at the top left-hand corner. Others expect work to be submitted via the course dropbox, with no hard copy required at all. (Document management systems make it easy to submit assignments electronically, but never send work that your instructors have to download individually unless you have been asked to do so.) As always, it's the audience that determines the format of the document. Therefore, whatever written work you submit, make sure it meets the expectations of each person who will be judging it.

Chapter Checklist

- ☐ Determine what document genre is called for according to your purpose for writing, your audience, and the context.
- ☐ Identify and complete all the expected sections, including references and appendices, as laid out in your table of contents.
- ☐ Format according to convention; then edit and proofread carefully.
- ☐ Arrange for binding and submission of a hard copy accompanied by a letter of transmittal.

Notes

1. IEEE. (2014) 2014 IEEE Taxonomy Version 1.0. [Online]. Available: http://www.ieee.org/documents/taxonomy_v101.pdf
2. LaTeX Project. (2014, Feb. 4). "LaTeX: A document preparation system." [Online]. Available: http://www.latex-project.org

Working Collaboratively

6

Chapter Objectives

- Assigning responsibilities for group projects
- Keeping the project logbook
- Collaborating on the project report
- Preparing to present your work as a group

Introduction

As you know, a large part of engineering depends on teamwork. It makes sense, therefore, to have a straightforward plan for collaborating with others in the design and execution of work that you do together as a team.

School offers many opportunities to develop teamwork skills in group projects. Your instructors have two main ways of evaluating these projects. Most often, the instructor assigns a grade for the work as a whole, and every participating student receives the same mark. If the team is unbalanced in terms of ability and experience, such weighting can mean that stronger team members end up doing extra work to make the project successful. If, on the other hand, the instructor assigns individual grades to each member according to the work he or she is deemed to have done, then it's even more important to make sure your contribution counts. Whether you have been designated to a team or whether you are allowed to choose collaborators with whom you have worked before, your instructors expect you to take advantage of this learning experience to broaden your skill base as you work together.

A number of roles exist in a team project, depending on how many people are assigned to a group, from keeper of the project log to slideshow planner. If this is the first time you've worked together as a team, spend time in your initial meeting getting to know the other members and discovering each person's strengths and weaknesses. One of you might be a good problem solver;

another might be a confident writer and editor. Take advantage of the opportunity to share your skills with the others. Everyone on a team has different abilities, so it's worth identifying these talents right from the start so that you can all benefit from the group experience.

Especially if there are more than three of you working on an assignment, you should begin by designating one person to head the team, someone who is both a good organizer and a natural leader. This person acts, essentially, as coach and captain to coordinate the team's efforts, to help assign tasks, assess progress, and generally oversee the presentation of the final product. Even if your team is being supervised by a faculty member, you should still select someone for the role of project manager. The adviser may offer suggestions, but you are ultimately responsible for the work you do as a team. The purpose of the exercise, of course, is to give you practical experience that will apply to your professional career.

One of the greatest worries associated with collaborative work is that a team will end up with one member who doesn't pull his or her weight. As a group, you need to make it clear right from the first meeting that although the project leader has overall authority for maintaining the momentum of the project, each person has responsibilities, too. Decide as a group what to do if someone misses a deadline or doesn't measure up. This is the time to establish the ground rules for contributions.

Whether or not your project will be presented in person—in class, at a conference, or in a boardroom—you will likely produce text in writing as well as a slideshow. Here are the steps to take to prepare and present work involving several people.

Getting Organized

As soon as you know who's in your group, plan a get-to-know-you session where you accomplish the following:

- Exchange contact information and set up an electronic home base for your group using Google Docs or the wiki available on your course's learning management system.
- Assign roles and responsibilities. Recognizing that everyone has a role to play is a key to successful teamwork. Whether it is a position, such as project manager, or a responsibility, such as keeping a list of expenses, make sure everyone knows and feels comfortable with what is

expected. Remember to appoint someone to be in charge of the project logbook; this person's job begins immediately.

- Define the nature and limits of your project. This is, for all intents and purposes, a needs analysis.
- Organize your project, both in terms of what has to be done (the plan) and when it has to be done (the schedule). Determine the optimal order of activities and decide which team members will be accountable for these. (If you have already studied the Critical Path Method in your engineering courses, use it to prepare a plan for the project [see p. 133].) It also makes sense to prepare a Gantt chart to identify timelines and responsibilities (see Fig. 6.1). The horizontal axis divides into the days, weeks, or months for the project; the vertical axis lays out the project stages and names the team member each involves.
- Establish an agenda for the next meeting (real or virtual) that calls for each member to give a progress report.

Your engineering textbook gives you strategies for undertaking a project. The key to being successful in collaborative work is how you chronicle your team's progress and manage the schedule. Be sure to allow enough time at the end for a full write-up, including plenty of time for the drafting of sections and the overall editing, proofreading, and formatting called for in a formal report. If you build this schedule into your Gantt chart and designate a project editor, you prepare everyone to contribute to the writing right from the beginning of the project. If you are making a public presentation, allow time as well for

Activities	Week 1	Week 2	Week 3	Week 4
Project Definition (All)	▰	▪		
Research (Mo, Andy)	▰▰			
Review Meetings (All)	▪	▪	▪	▪
Prototype Development/Testing (Kim, Faye)		▰	▰	
Report Writing (Kim, Andy)		▰▰	▰	
Preparation for Presentation (All)			▰	▰

Figure 6.1 Gantt project planning chart

the creation of a slideshow that puts your work in the best light. Two essential parts of the process are the *project logbook* and the *planning templates*.

Maintaining the Project Logbook

Like the official minutes of board meetings, material included in your engineering project logbook [1] establishes a legal record. In keeping track of the ideas you've had and progress you've made, the logbook offers proof of originality (intellectual property) and helps others retrace your steps. That's why it's so important to have members of your team corroborate its contents by initialling and dating each entry.

Despite the practicality, reliability, and universality of computers, the project logbook has a hard cover and sewn-in pages. If the pages aren't already numbered, do this at your initial meeting when you designate someone on your team to be the keeper of the log. Of course, anyone in your group can make an entry, but it's a good idea to have one person responsible for overseeing what gets recorded.

For legal reasons, it's important to write in non-erasable ink (never pencil) and to use a colour that scans well, such as black or red. The following are additional conventions that help you legitimize the contents of your logbook:

- Use the logbook to record, in chronological order, all your activities as a team—your meetings, discussions, ideas, decisions, sketches, designs, calculations, solutions. For each entry, record the date and time as well as the names and contributions of everyone in attendance. If you haven't worked on the project for a while, make an entry about the pause in activity, stating the reason and having the entry initialled by team members.
- Use the regular space available on each page to make your entries without leaving spaces or extraordinary margins. (You don't want to leave yourselves open to the accusation that you are planning to add to an old entry at a later date.) It's normal to fill all pages, but if you decide to start each new entry on a fresh page, draw a diagonal line neatly through the space so that it cannot be used again.
- Keep the order of entries chronological and dated. If you want to add to an earlier entry, it is standard practice to add the material in chronological order at the end of your entries and make clear cross-references to the earlier pages and dates. Initial all of your cross-references well.

Of course, if the addition is not much more than a phrase or a clarification, you can add it to the original entry in a different colour of ink—just be sure to date and initial it.

- If you want to include photographs, spreadsheets, or other pieces of paper, glue them in neatly, initial them, and date them.
- Be as objective and accurate as possible in recording discussions among team members. Always give reasons for decisions, and use wording that cannot be misinterpreted by anyone reading the logbook.
- If you want to delete something, either because it is a mistake or because it is not relevant, cross it out with a single tidy line, so that you can still read what it says. Do not tear out pages, erase entries, or use correction fluid. These actions reduce the value of your project notebook, and your team loses credibility.

These precautions are nothing less than standard practice in the engineering world, so it's worth getting used to following them while you are still a student. In fact, if you are working on a groundbreaking project, your logbook takes on even more importance. As a permanent chronological record of all the work your team has done, it may even be required as evidence in court. To add to its authenticity, you may decide to have it regularly corroborated by independent witnesses—especially if your work is precedent-setting.

As you can imagine, it's hard to find knowledgeable people willing to read and sign every entry in a logbook. They must also be disinterested third parties with no reason to make false attestations about your work. Still, if your team is serious about proving ownership of original material, getting professional corroboration is worth the effort. Have your witnesses verify each entry within hours of its writing. Use the wording shown in Fig. 6.2, and be sure your project leader signs as well.

The preceding entry was made by _____ on this ____ day of _____, 201__.

Project leader: _____

Witnessed and understood by _____ on this ____ day of _____, 201__ and by _____ on this ____ day of _____, 201__.

Figure 6.2 Testimonials

Preparing the Project Report

As you learned in Chapter 5, the report is the printed culmination of your work on a project. When you are working as a member of a team, the preparation of this document takes on new dimensions, since the contributions of several people have to be synthesized in a coherent document. It's a good idea to start by assigning the job of master compiler/editor to one of the team members, the person who has the most experience with writing.

In delegating tasks, try to make the burden of work as fair as possible—even though fairness is an elusive goal. Most of the complaints about working collaboratively are that one person has done more work than others, or has done a lot less, even though everyone gets the same grade. This is no different from most team sports, where athletes recognize the advantages of collaboration despite differences in talent and experience. The strength of a team, after all, depends on the combined power of individual contributions.

Arrange to have everyone on the team get together at a strategy meeting near the end of your project. Try to leave yourselves about a week to prepare the final report. Even if you have ample material to draw from, you should never underestimate the effort it will take to write it up.

Together, use your project logbook to establish a comprehensive table of contents outline for the final report, and assign writing responsibilities to team members according to their expertise in the project. The work will be much easier if you prepare *planning templates*. Use the storyboard model in Fig. 6.3 and have team members create a file for every section or subsection that they are responsible for, including appendices. Together with everyone on the team, the project editor checks the numbering against the outline to ensure that all parts of the project have been assigned and there are no gaps in the organization.

Drawing up plans is essentially the same work that story developers do in preparing scripts for production. It's their way of making sure that there are no gaps in the plot.

If you are the designated editor, here's how you use the storyboarding method to prepare the final draft of your group's project:

- Set firm deadlines for submissions. It's ideal if you can build in a little wiggle room to allow for unexpected delays.
- Check the completeness of all submissions against the table of contents. This is the best way for you to identify what material, if any, is outstanding. (Follow up immediately with your contributors if you discover anything

Section/subsection number: _____ [position in Table of Contents]

Heading: _____ [title or subtitle]

Team member: _____ [name of person responsible for the draft]

E-mail:_____

Purpose: _____

Organizational pattern: _____

Key points:

-
-
-
-

Graphics, support [tables, graphs, diagrams, equations, etc.]:

-
-
-

Figure 6.3 Storyboard planning chart

missing.) More importantly, you will know which sections are complete, so that you can start your editing work there.

- Use the format rules for the project as guidelines for editing and standardizing the text.
- If you are required to write a reflection paper on your participation in a group project, follow a standard chronological pattern for your summary of work completed and then elaborate (reflect) on what you learned from the experience. Team members are normally expected to complete such reflections individually, but if you are required to prepare one as a group, follow the same writing and editorial process as you did for your project.
- If there will be a public presentation, meet as a group to assign responsibility for putting together the slideshow.
- Reserve time to do one last thorough editing in a single sitting. By working your way from beginning to end of the draft, you will be more likely to discover and eliminate inconsistencies. By the end of this session, you should be confident that the style and tone are consistent

throughout, that the text is readable, that the grammar and punctuation are accurate, and that the charts and graphs are consistent and understandable.

- Once you are satisfied with your editing, prepare the abstract or executive summary to be included at the beginning of the report. It is also your responsibility as editor to write the letter of transmittal (see p. 54).

Now you are ready for one last project meeting. Make sure it is at least a day before the final project is due. Everyone on the team should already have proofread the report and its cover letter. If there is a slideshow, work through it all together too. The more proofreaders, the more likely it is that someone will catch an error. As the editor, you are in charge of making last-minute changes or corrections, but it's still important to share the overall responsibility for spotting them. Once everyone has approved the final text, produce a presentation copy for submission.

If your team is going to present the project to an audience, match the text to your slideshow by cutting and pasting the final draft of the paper to make a script for each member of the group. Take the time to rehearse at least twice so that you can verify and refine the timing. You will find guidelines for individual presenters in the following chapter.

Chapter Checklist

☐ As soon as you are assigned a group project, schedule a brainstorming session where you assign responsibilities and exchange contact information.

☐ Establish and maintain a project logbook as a professional record of the work you do together.

☐ Take advantage of storyboard planning templates to ensure you have been thorough and comprehensive.

☐ Have all team members approve the final report and/or slideshow. Rehearse the presentation at least twice.

Note

1. Peter H. Roe, et al., "Project logbook notes," [handout], SyDe 461/462, University of Waterloo, 2008.

Giving Presentations

7

Introduction

Engineers are expected to be able to present research, solutions, creative ideas, and proposals not just in writing but also in person. To help you prepare for professional life, many of your courses include class presentations as part of their requirements. Whether you are asked to report on work in progress or to present a finished project, the key to success is careful planning. Even if public speaking is not one of your natural talents, you can still present material effectively as long as you are well prepared and well organized.

Consider the characteristics of classes or seminars you have enjoyed. What made them so good was likely the confidence, know-how, and enthusiasm of the speaker. Most audiences prefer presenters who seem well prepared, who speak without reading from notes, who use a range of relevant visual aids, and who look animated and interested in what they are talking about. Still, it's not easy to stand up in front of a group of people, no matter how well prepared you might be. To make the best impression, follow these guidelines.

Preparing Your Presentation

Understand your purpose

When preparing for any presentation, remember that you know more about your specific focus than anyone else in the room (including your supervisor). Your aim is to convince your audience that you know what you're talking about and that your grasp of the subject matter goes beyond what you are including in your presentation. Prepare yourself for questions by doing the necessary background reading. (For example, you need to be familiar with all the resources you list in your bibliography.) If you are presenting a report on work in progress, develop questions to ask your audience so that you can take advantage of their ideas. If you are delivering a final project report, be ready to justify your recommendations by producing support and evidence from other similar projects.

Consider your audience

Your primary consideration must always be your audience. To appreciate their information needs, you should identify exactly who will be attending your presentation. Faculty members? Fellow students? Your project supervisor? A potential employer? Knowing how large your audience is and whom it includes helps you determine both your approach and your tone. After all, there are distinct advantages to addressing people in person:

- You can establish a rapport with your audience by personalizing your focus with references to *you*, to *I*, and to *we*. Your language and approach will be more casual, comfortable, and natural than the language expected in a formal paper.
- You can count on your audience to give you immediate feedback. If people look puzzled or unconvinced, you can take the time to restate your ideas or expand upon them in a way that is not possible in writing.
- You don't have to worry about such conventions as spelling, formatting, and punctuation (except as they apply to your visual aids) because you are speaking, not writing.
- You can use your voice, facial expressions, and body language to add energy to what you have to say.

The most important thing to remember is that your audience has a reason for being in attendance. Rather than concentrating on how you feel about speaking publicly, focus your attention on delivering the interesting, new information your audience expects.

There are two sides to the question of whether you should provide an audience with a handout featuring copies of your slides. Some believe that the reproduction of slides gives a lasting reminder of contents, which is useful particularly if the material might appear on a test. Others argue that it's both a distraction and a waste of paper. Base your decision on the purpose of your presentation. It is easy to make your slides available after the event, either on a course website or by e-mail, so include your e-mail address on the slide that remains visible during the question period at the end of your presentation.

Your audience will appreciate having an outline—as long as it contains specifics (something more than *Introduction/Discussion/Conclusion*). If you do provide a handout, put your name, e-mail address, title, date, and course number at the top, and include a bibliography (perhaps annotated, see p. 36) and additional references at the end. If possible, leave room for your listeners to take notes as you talk. They will also be grateful for written definitions of terminology that might not be common knowledge.

Plan your presentation

Except for very formal presentations of conference papers, never simply read a prepared text out loud. (Unless you are a skilled public speaker it's hard to keep such a presentation sounding lively.) That isn't to say that you shouldn't write it all out first—doing so will help you be better organized and perhaps less nervous. But if you have the full text in front of you at the presentation, the temptation to read from it can be overpowering. It's better to be prepared to speak off the cuff.

Use your slides to help keep you organized. As you develop your series of slides you are essentially planning the presentation. Take note of the special demands for each section:

1. **Introduction.** Establish the context for your presentation by relating it to topics you and your classmates or colleagues have already discussed as a group. Start by posing a question to be answered, or by making a provocative statement to catch the audience's attention and to identify your subject.
2. **Discussion.** Follow a clear order, just as you would in writing (see pp. 7–8). Include a title or caption on each slide and take advantage of bullets to list specifics.
3. **Conclusion.** Use the final minutes of the presentation to summarize and reinforce your main points. Bring the presentation to a satisfying conclusion for your audience by referring to the question or issue raised in your introduction.

While you do your planning, keep in mind how much time you have been allotted. It's essential to stay within the time limit. (You can easily recall an occasion when you've been annoyed watching someone else go on longer than expected.) Ideally, you should plan to make your talk a little shorter than the amount of time you allow for questions. As a rough guide, it takes a minute or two to read a page of double-spaced text; time yourself accordingly by rehearsing with a written-out copy of your presentation. Use your text to prepare your slideshow, and take advantage of the notes feature if you need prompts while presenting.

Use effective visual support

There are good reasons to take the time to develop visual aids to accompany your presentation. First, they attract and focus the audience's attention. If you become self-conscious in front of a group, you will be more at ease when the eyes are on your visual aids rather than on you. Second, they allow you the flexibility to present your information in a variety of ways, for maximum effect.

With the availability of interactive whiteboards and digital projectors, your ability to use visual aids in a presentation is limited only by your own ingenuity. You must be absolutely sure, however, that your visuals do not take away from your content. Novice presenters can get so carried away by the potential of technology that they defeat their purpose. No matter how appealing the bells and whistles seem, remember that you are making a presentation, not a cartoon.

A slideshow that includes video clips and sound as well as animated diagrams is effective only if these elements are necessary and appropriate. Instructors commonly complain that the time taken to design extravagant visuals would have been better spent on research. Don't leave yourself open to this criticism: work to make professional slides that enhance the content rather than call attention to their excesses.

The following suggestions apply to any material you might want to display during a presentation.

Keep things plain and simple

The demand for clarity and simplicity governs every aspect of visual aids. It is better to put too little information on a slide than too much. Your slides should help you make your point, not detract from it. The following suggestions will help you keep things simple:

- **Use a plain font.** Any word processing or graphics software includes dozens of font options. Stay with the clean and simple ones like Courier or Arial, reserving Comic Sans for informal occasions only. Avoid all fancy fonts, including italics. Use bold face instead if you want to emphasize something.

- **Choose an appropriate font size.** Verify that your slides can be read from the back of the room in which you will be giving the presentation. Remember that the size of any type you use will depend on how far your projector is from the screen, not on the font size you used to prepare the slide. If you keep to a font size from 20 to 36, you should be fine.

- **Use a simple background.** No matter how tempting it is to try out the variety of colourful bullets, backgrounds, and borders, choose a plain, light-coloured background, and use the same one on every slide. If your school or department has a standard template, use it.

- **Don't overuse colour or other effects.** Using fonts of different colours can provide effective contrast, but using too many is distracting. Avoid flashy effects, which get tiresome for the audience. It's worth noting that a simple illustration in black and white is often easier to interpret than a photograph. Slideshow software offers the option of introducing text in a variety of ways, even letter by letter. Don't be tempted by exotic possibilities. Use the screen to display everything that relates to a single point at the same time. It is annoying for an audience to have information displayed in tiny portions. No one likes to be kept in suspense.

- **Don't pack too much information onto one slide.** If you treat your slides as your script, you can be tempted to read directly from them. Instead, use bullets as prompts to allow yourself the time to elaborate as you talk. Whenever you have a table, diagram, or graph, use the simplest, cleanest version possible and explain it thoroughly to your audience. If you are standing right next to the screen, use broad gestures to point to relevant parts of your slides (it gives you something to do with your hands). Laser pointers or an electronic mouse, while popular, tend to bounce around annoyingly, so it's better to avoid them unless you are very practised. If the screen is too large, consider what animations you can add that will draw attention to parts of text that you'd want to point to.

- **Don't offer too many visuals.** The number of slides you include depends to some degree on your topic and the kind of material you are presenting. A rule of thumb for determining how many slides to use during your presentation is to plan for no more than one slide every two minutes. You'll go through some quickly, but you'll take more time with others. Be sure that you include only slides that you have time to discuss. There's nothing worse than rushing through slides. Your audience notices immediately that your timing is off.

Keep things organized

The second rule for visuals is to make sure your material is well organized. If you follow a consistent pattern of presentation, the audience is able to follow along more easily. At the same time, you will be more comfortable making your presentation. The following suggestions will help you keep your slides and your presentation organized.

- **Begin with a title slide.** A title slide orients your audience. It should contain the title of the presentation, your name, the date, and perhaps the name of the course. Have your title slide on the screen as people enter. Wait for everyone to be seated before you begin to speak.
- **Have an overview slide.** Put an overview of your presentation on the screen as you introduce your topic. This has the advantage of helping your audience estimate your timing. Do not, therefore, include so many points that your audience fears you will never get finished.
- **Use headings and subheadings.** Most of your slides should be in point form, using numbers or bullets, with headings and subheadings. If you do this, the audience will be able to distinguish between the main points and the elaborations. Remember to be consistent: if you're using numbers, don't switch needlessly between Arabic and Roman numerals; if you're using bullets, use the same style throughout.
- **Consider section breaks.** If you are giving an extended presentation that lasts 30 minutes or more, consider dividing your presentation into sections, with separate title slides for each section. Anything that helps the audience to follow the direction of your presentation is worthwhile.
- **End with a summary and/or conclusions slide.** If you have been discussing an ongoing project, list the questions you want to pose to the audience at the end. If you are making a proposal, your final slide should identify *next steps* (your suggestions for follow-up or future planning). Avoid having a last slide that features the words "Questions?" or "Thank you!" A summary of your main points is much more useful. Do include your e-mail address on the last slide though, along with your website URL if you have one.

Rehearse your presentation

Once you have prepared your slides, take the time to run through your presentation at least once. See if it's possible to book the actual room to rehearse in. If so, take the opportunity to check the equipment and the readability of your visual aids while you get used to standing at the front. It's best if you can rehearse with someone whose opinion you trust. Take advantage of his or her feedback to improve your delivery. A couple of dry runs will help you identify areas you need to strengthen and let you know if you are keeping to your time limit. Keep the following points in mind as you rehearse.

- **Speak slowly and deliberately.** Make sure your voice can be heard at the back of the room. Remember that a room full of people will muffle sound, so you have to talk even louder at the official presentation.
- **Practise looking around the room as you talk.** Choose three focal points where you will direct your gaze during your presentation—left, right, and centre. By looking from one place to the next, you appear to include everyone. That way you seem less wooden and more comfortable addressing your audience.
- **Time your delivery accurately.** Be prepared to cut out sections rather than go over time. In fact, as you rehearse, mark places where you could shorten your discussion if necessary. Then, during the presentation, you can just move to the next section smoothly as you keep to your time limit. After your rehearsal, be ruthless about editing slides you don't need. Your audience won't even notice what you've left out.

Making Your Presentation

Dress comfortably

Dressing comfortably means dressing for the occasion, not overdressing. Don't wear anything that you feel awkward in, including new shoes. There's no need for a business suit unless you know the rest of the audience will be wearing one too. A good rule of thumb is to dress the way your instructors and TAs do. Something that's clean, neat, and casual sets the right tone.

Leave yourself plenty of time to get there

Don't arrive at the last minute. Leave yourself time to set things up and make sure all the equipment works. There's nothing more disconcerting for you or your audience than to have a projector fail or a slideshow go missing. If everything is ready before you're scheduled to begin, you can take time to relax. Best of all, you won't get flustered rushing to set things up with your audience looking on.

If, in a worst-case scenario, there is a complete equipment failure, take advantage of whatever visual aids are available—blackboard, white board, flip chart, or overhead projector. The well-prepared presenter has a hard copy and a set of overheads just in case the equipment malfunctions.

Speak in a calm, clear voice

When you speak, be sure you're loud enough that everyone in the classroom can hear you. Also, try to put some energy into what you are saying. (It is difficult to remain attentive to even the most interesting presentation delivered in a monotone.) One of the tricks of professional speakers is careful breathing. Taking a silent, deep, slow breath before you begin is a way to master nervousness. (It's like counting to 10 to avoid losing your temper.) Breathing slowly and deliberately while you deliver your presentation not only gives the impression that you are in control but relaxes your audience.

Don't apologize

Never start with excuses: "You'll have to forgive me. I'm really nervous about this" or "I sure hope this equipment is going to work properly." Audiences are more receptive if you focus on your topic rather than on your nervousness. After all, you don't want to suggest that you are unsure of your material. Most of your audience won't even notice your discomfort—as long as you don't draw attention to yourself by apologizing.

If you are someone who finds it difficult to stand up in front of a crowd, consider options for confronting your discomfort. Many colleges and universities offer courses in public speaking, and some local organizations even host regular meetings of Toastmasters International. If your future involves regular presentations and you are naturally reticent, consider getting some practice to make yourself more comfortable. Start by checking with the career services department on your campus.

Maintain eye contact with your audience

Look around the room as you speak, shifting your gaze from one to the other of your focal points. When you look at people, you involve them in what you are saying. As you scan the faces in front of you, you can watch for signs that your audience is following. If you sense confusion, adjust your talk accordingly by explaining a difficult point or slowing your pace.

Work with your visual aids

Take full advantage of your slides (while remembering that the visual material should enhance your presentation, not deliver it for you). Here are some guidelines for using visual aids effectively:

- If you are using your own laptop, remember to turn off pop-up features and audio to prevent incoming messages from interfering with your presentation.
- When referring to a point on a slide, use different words to elaborate on what appears there. Use your slide as a set of notes to expand on. Don't simply read the words on the screen.
- Be sure to leave your audience enough time to make sense of each visual. There is nothing more frustrating than watching slides flash by without having a chance to take them in. It is better to delete a visual than to rush it by your audience. This is another reason why it's better to have only a few uncluttered, uncrowded slides. Audiences need time to understand something new. You don't want people to become frustrated and negative because they feel rushed.
- Explain your figures. If it's a graph, make it clear what the x- and y-axes represent, then explain what the graph shows. If it's a diagram, take the audience through it step by step. Remember that you are much more familiar with the material than your audience is. Never take for granted that anything is so obvious that it doesn't need to be explained.
- During your presentation, move around periodically so that you are not permanently blocking anyone from seeing the screen. Avoid putting yourself in front of the projector: it's distracting to an audience to see a presenter lit up by text from a slide. Staying at the side of the screen or board allows you to point effectively to important material while guiding your audience's eye.

Pace yourself

As you speak, adjust your speed to your content. If you're discussing background information that everyone knows from class, you can go a little faster. If you're describing something complex or new, slow down. It often helps to explain a complicated point a couple of times in slightly different ways. Don't be afraid to ask your audience if they understand. Almost certainly, someone will speak up if there is a problem. But don't let this rattle you. A one-sentence answer will allow you to continue on schedule.

Monitor your time

Stand so that you can see a clock throughout your presentation, but don't make a big deal of positioning your smartphone or watch at the beginning. (Many of those who make elaborate gestures like this go over time anyway.) If you've carefully rehearsed your presentation, you should know roughly how long it will take. Always leave extra time for questions that people might ask.

End confidently

Bring your presentation to a strong finish by summarizing the main points you have made and drawing conclusions. Remember to have these available on a concluding slide so that they can be left there for the ensuing discussion. If you raise questions in your conclusions, you can use them to help structure the question period to follow.

Be prepared for questions

The question period is the time when you can make the best impression. It is an opportunity for you to demonstrate your thorough understanding of the topic and even to reinforce one or two points that you think you may have missed. If you know your material well, you should have no problem dealing with the content of the questions, but the way you answer these questions is important, too:

- Don't introduce the question period with a hurried "Any questions?" Such an abrupt approach makes you seem anxious to rush through or altogether dodge the finale. You appear sincere about welcoming questions if you phrase your request slowly this way: "Now, what questions do you have?"
- It's a good idea to repeat a question for the audience, especially if you are in a large room where everyone may not have heard it. Repeating the question can clarify it for you and buy you time to think before you answer.
- If you didn't quite hear or didn't understand a question, don't be afraid to ask the person who asked it to repeat or clarify it.

- Keep answers short and to the point. Rambling is neither helpful nor convincing. Long answers often confuse the audience, so it's best to start briefly and then elaborate if necessary.
- If you don't have an answer, say so. It's also okay to ask your audience for suggestions or refer the question to the course instructor. Certainly, it's better to admit that you don't know an answer than to guess or to make up a response that everyone knows is not correct.

Whether you are speaking alone or as part of a team, you have the best chance of success if you take your time to prepare, rehearse, and refine your presentation. Take advantage of opportunities to gain experience speaking in public, and your efforts will pay off in increased confidence and control.

Chapter Checklist

- ☐ Plan your presentation with your purpose, your topic, your audience, and your time limit in mind.
- ☐ Keep the information on your slides clean and simple.
- ☐ Time your delivery carefully, and rehearse until you are comfortable with your material.
- ☐ Treat your audience as collaborators to share ideas with.

8

Writing in Other Contexts

Chapter Objectives

- Managing the requirements of non-technical writing assignments
- Organizing unified introductions and conclusions for essays and reports
- Writing commentaries on books and articles

Introduction

Not all the writing you do is scientifically oriented. In addition to the summaries and reports you prepare in technical contexts, you may find yourself required to write essays, discussion papers, or reviews. Fortunately, your engineering training helps you adapt traditional organizational approaches to these different writing projects.

Preparing to Write

If you are one of the many students who dread writing, you will find that following careful steps in planning and organizing makes the task easier—and the result better.

Some people claim they can write without any planning at all. On the rare occasions when they succeed, their writing is usually not as spontaneous as it seems; in fact, they have thought or talked a good deal about the subject in advance and have come to the task with some ready-made ideas. More often, students who try to write a lengthy essay without planning just end up frustrated. They get stuck in the middle and don't know how to finish, or they suddenly realize that they're rambling.

Most writers agree that the planning stage is the most important part of the whole process. Certainly the evidence shows that poor planning usually leads to disorganized writing. In much student writing, the single greatest improvement would not be better research or better grammar but better organization.

This insistence on planning doesn't rule out exploratory writing (see p. 255). Many people find that the act of writing itself is the best way to generate ideas or overcome writer's block; the hard decisions about organization come after they've developed some text. Whether you organize before or after you begin to write, however, at some point you do need to plan.

Narrow your focus

Some instructors provide a range of topics to choose from; others consider it part of the assignment to limit and develop a subject. If you don't have a specifically assigned subject, begin by identifying an area that you know something about, one that you like and don't mind writing about. Ask questions that help you gain some focus. Your aim at this point is to frame the single-sentence statement—the thesis—that sets the stage for the discussion in your paper.

One scheme for applying systematic question strategies is the three-C approach. It makes you look at a subject from three different perspectives, asking basic questions about *components*, *change*, and *context*:

Components
- Into what parts or categories can the subject be broken down?
- Can the main divisions be subdivided?

Change
- What features have changed?
- Is there a trend?
- What caused the change?
- What are the results of the change?

Context
- What is the larger issue surrounding the subject?
- To what tradition or school of thought does the subject belong?
- How is the subject similar to, and different from, related subjects?

What are the components of the subject?
In other words, how might you break the subject down into smaller elements? This question relates to *classification* as a method of development. It forces

you to look closely at the subject while avoiding oversimplification or easy generalization.

Suppose that your assignment is to discuss networks. After asking yourself about components, you might decide to split the subject into (1) public networks and (2) private networks. Alternatively, you might focus your discussion into priorities of network design: (1) accessibility, (2) security, (3) scalability, and (4) budget. If these components seem too broad, you might break them down further, perhaps examining individual providers.

Similarly, if video games were a topic, you could ask, "What are the different genres?" (for example, driving simulations, adventure games, and real-time strategy games). Or you could sort video games according to their origins by asking, "What are the tabletop sources of these games?" (for instance, board games, puzzles, or cards). If you were interested in looking at stock market analysis, you might ask, "What relative risk is associated with each of three common trend predictors?"

Approaching your subject this way helps you appreciate its complexity while avoiding broad generalizations. Asking questions about the components of your subject may help you find one element that is not too large to explore in detail.

What features of the subject suggest change?

This question helps you to think about trends. Consider it an extension of the traditional chronological order patterns you are used to in engineering. It points to antecedents or causes of an occurrence as well as the likely results or implications of a change.

Suppose you have decided to focus on communications systems. You might consider whether the proliferation of smartphone providers has had an effect on spending for research and development. You might look at increases in traffic congestion on company networks.

For a paper on video games, you might examine increases or decreases in popularity of a type of game, or trace improvements and developments in general. You could even track or account for complaints of how preoccupation with games inhibits people's social skills.

In terms of the stock market, you might ask, "Have there been changes or trends in the rate of criminal offences committed?" or "Have there been changes in the nature or definition of insider trading?" Then ask, "What are the causes or results of these changes?"

What is the context *of this subject?*

Into what particular school of thought or tradition does the subject fit? What are the similarities and differences between this subject and related ones? The following are typical context questions:

- How do Google Chrome and Mozilla Firefox compare?
- What are the essential differences between two databases?
- What characteristics do the most popular video games share?
- How predictable is a typical buy-sell algorithm in a bull or bear market?

All these questions lead to better questions and answers from which you refine your topic and determine your controlling idea.

Develop the thesis

All writing needs a controlling idea around which the material can be organized. This central idea is usually called a *thesis*, though in expository writing you may prefer to think of it as a *theme*. Consider these statements:

THEME: There are three standard techniques for treating contaminated groundwater.

THESIS: Governments should force polluters to use bioremediation to treat contaminated groundwater.

The first is a straightforward statement of fact. A paper centred on such a theme would follow a classification structure, describing the three remediation techniques one at a time. By contrast, the second statement is one with which readers might well disagree. A paper based on this thesis would have to present a convincing argument. The expository approach leads to an informative and interesting paper, but the persuasive approach is more likely to produce strong writing—as long as there's solid support for the argument. The key thing to remember about the theme or thesis, however, is that you must be able to state it in a single sentence that is limited and precise. Ideally, this sentence should prepare the reader for the discussion to follow. It is, in other words, what holds a paper together. Measure the simplicity and strength of your thesis by making sure it fits into the following pattern: It is true that _____.

Create an outline

Individual writers differ in their need for a formal plan. Some say they never have an outline; others maintain they can't write without one. Most fall somewhere in between. Since organization is so vital to the success of a paper, it's worth knowing how to draw up an effective plan. Of course, the exact form it takes will depend on the pattern you use to develop your ideas—whether you are describing, classifying, or comparing, for example (see pp. 7–8).

For most students, an informal but well-organized outline in point form is the most useful model. As you've seen in earlier chapters, your table of contents will sometimes be all you need before you start to write.

The following is an example of an outline:

INTRODUCTION: Definition

THESIS: Virtual environment technologies have tremendous potential in the health care field.

I. Tools for Virtual Environment Applications
  A. Head-Mounted Displays
  B. Instrumented Clothing
  C. Spatialized Sound
  D. Other Interface Technology
      1. Virtual balance sensors
      2. Virtual eyes

II. Virtual Environment Applications in Medicine
  A. The Virtual Medical Classroom
  B. Virtual Reality–Assisted Surgery
      1. Preview
      2. Simulation and rehearsal
      3. Operation
  C. Virtual Rehabilitation

CONCLUSION: Design Implications

You can take advantage of the same formula to organize yourself. The guidelines for such an outline are straightforward:

- **Code your categories.** Use different sets of markings to establish the relative importance of your entries. Most computer outlining applications provide default coding but permit alternatives.

- **Categorize according to importance.** Make sure that only items of equal value are put in equivalent categories. Give major points more weight than minor ones.
- **Use parallel wording.** Phrase each entry in a similar way to make it easier to be consistent in your presentation. (See p. 200.)
- **Check lines of connection.** Make sure that each of the main categories is directly linked to your central thesis; then see that each subcategory is directly linked to the larger category that contains it. Checking these lines of connection is the best way of confirming that your paper is well organized.
- **Be consistent.** In arranging your points, avoid discrepancies or contradictions. You may choose to move from the most important point to the least important, or vice versa, as long as your organization is predictable.
- **Be logical.** In addition to checking for lines of connection and organizational consistency, make sure that the overall development of your work is logical. Does each heading/idea/discussion flow into the next, leading your reader through the material in the most logical manner?

Be prepared to adapt the structure at any time in the writing process. Your initial outline is not meant to put an iron clamp on your thinking but to relieve anxiety about where you're heading. A careful outline prevents frustration and dead ends—the feeling of "I'm stuck, and where do I go from here?" But since the very act of writing usually generates new ideas, you should be ready to modify your original plan. Just remember that any modifications must have the consistency and clear connections required to maintain unity.

Writing an Essay

One major difference between reports and essays is the format. Reports are organized around headings and subheadings, which help orient both the writer and the reader. Papers and essays, on the other hand, depend entirely on the writing to provide unity and coherence. Essay writing sometimes seems complicated to do without signposts.

Many essay writers find it easier to compose a first draft as quickly as possible and do extensive revisions later. However you begin, never expect the first draft to be the final copy. You already know that revising is a necessary part of the writing process and that careful revisions make the difference between mediocre writing and good writing.

If you ever face writer's block, remember that you don't need to write all parts of your essay in order. In fact, many students find the introduction the hardest part to write. If you face the first blank page with a growing sense of paralysis, leave the introduction until later and start somewhere in the middle with a subtopic you know well. Once the body of the essay is fleshed out, your introduction is easier to write. Many experienced writers, not only those with writer's block, find this a productive way to proceed.

Develop the introduction

The beginning of an essay has a dual purpose: to indicate your topic and approach and to whet your reader's appetite for what you have to say. One effective way of introducing a topic is to place it in a context—to supply a kind of backdrop that puts it in perspective. Your structure can follow the **GEST** pattern below:

GENERAL STATEMENT (**G**): *N* is the larger subject.

EVIDENCE (**E**): Here are some applications/examples.

SUMMARY STATEMENT (**S**): These applications/examples include *X*.

THESIS (**T**): *X* is the focus of this essay.

Sheridan Baker [1] calls this the funnel approach because it works its way from broad to narrow in a few short sentences. The funnel introduction works for almost any kind of essay. The following example sets the stage for a discussion of needs analysis in design work:

(**G**) Most products or structures eventually wear out or become obsolete. (**E**) Sometimes they can be converted to other uses (the horse barn becomes a garage and the icebox becomes a storage cabinet), or they are scrapped and the materials reused. (**S**) In each case, the environment in which a design must be converted or scrapped has a serious effect on the problem statement. By analyzing this salvage or conversion environment, the designer can determine the effective life for new products and structures. (**T**) This is the exercise known as a needs analysis.

In an essay, you try to catch your reader's interest right from the start. The following variations on the funnel pattern can help you do so:

- **The quotation.** This approach works especially well when the quotation is taken from the person or work that you will be discussing.

- **The question.** A thought-provoking question can make a strong opening. Just be sure that you do actually answer it by the end of your essay.
- **The anecdote.** This is the kind of concrete lead that journalists often use to grab their readers' attention. Save this approach for your least formal essays—and remember that the incident must really establish a setting for the ideas you are going to discuss.

However you choose to start a paper, your introduction must relate to your topic. Never sacrifice relevance for originality. Finally, whether your introduction is one paragraph or several, make sure that by the end of it your reader knows exactly what to expect from the paper that follows.

Develop the body

In an essay, it takes several paragraphs to develop an idea fully, and each new paragraph signals a change in the way you approach that idea. Skilled skim-readers know that they can get the general drift of a piece of writing simply by reading the first sentence of each paragraph. The reason is that most paragraphs begin by stating the central idea to be developed. If you are writing your essay from a formal plan, you will probably find that each section and subsection contains a topic sentence that indicates how the paragraph relates to the idea being developed.

The topic sentence does for a paragraph what the thesis sentence does for an essay: it's a statement introducing the point of the paragraph. It identifies, in a general way, the main idea to be developed. Often the organizational pattern (process, cause/effect, description, classification, comparison) is predictable (see pp. 7–8). The keywords in the topic sentence provide the foundation for the unity and coherence of the paragraph.

Like the thesis statement for the essay as a whole, the topic sentence is not obligatory: in some paragraphs, the controlling idea is stated near the middle or even the end; in others, it is merely implied. It's still a good idea to think out a simple topic sentence for every paragraph. That way you'll be sure that each paragraph has a readily graspable point and is clearly connected to what comes before and after. When revising, check that each paragraph has a topic sentence, either stated or implied. If you find that you can't formulate one, you should probably rework the paragraph.

SEES provides a basic pattern for a paragraph developing a topic in the body of an essay. It also supplies a useful model for the paragraph answers expected in definition questions on exams. SEES presents a general statement (topic sentence) followed by specific support:

TOPIC SENTENCE (S): *A* is the topic of this paragraph. (What's the point?)

ELABORATION (E): This is how it will be focused. (Can you be more specific?)

EVIDENCE/EXAMPLES (E): These are some applications/examples. (Can you prove it?)

SUMMARY (S): This is what it all means. (So what?)

In an essay, the summary sentence often establishes the connection between the paragraph and the main point of the essay. If this link is obvious, or if the next paragraph deals with a similar subject, the summary may not be needed. Still, the SEES pattern is useful for confirming that you've developed a point thoroughly. Many of the paragraphs in this book draw on this model, as does the one below:

(S) One example of varying missions, outputs, and inputs comes from considering the differences between North America and Britain when it comes to home heating. (E) Fireplaces perform very different functions in both places: (E) in North America they are aesthetically pleasing, and in Britain they are used for warmth. (S) Thus their missions and outputs differ, even though the inputs are the same. Of course, if the houses catch fire, the missions become the same—to put the fire out no matter what the country.

Maintain focus

A clear paragraph should contain only details that are in some way related to the central idea. Try structuring it so that all the details are seen to be related. One way of establishing these relations is to keep the same grammatical subject in most of the sentences that make up the paragraph. When the grammatical subject keeps shifting, a paragraph seems to have more than one topic sentence, as in the following example [2]:

orig. The speed of computer systems nowadays has led to many innovations. Artificial intelligence (AI) is an interdisciplinary sector of computer science. A lot of research is being done to recreate the cognitive processes of humans with computer software. You can already find some special applications with a limited number of possible decisions. Digital cameras use artificial intelligence to select the appropriate display window from three to ten choices for focusing the picture. People get a lot of entertainment and challenge out of computer games, where the artificial opponent has to make all the right decisions in order to beat the human.

In the example above, the grammatical subject (underlined) changes from sentence to sentence. Notice how much stronger the focus becomes when all the sentences have the same grammatical subject—either the same noun, a synonym, or a related pronoun:

rev. Artificial intelligence (AI) is an interdisciplinary sector of computer science. It deals with methods for recreating the cognitive processes of humans with computer software. To date, artificial intelligence has been used for applications with a limited number of possible decisions. In digital cameras, for example, artificial intelligence selects the appropriate display window from three to ten choices for focusing the picture. One of the most complicated AI applications is the computer game, where the artificial opponent has to make all the right decisions in order to beat the human.

Naturally it's not always possible to retain the same grammatical subject throughout a paragraph. If you were comparing two computer games, for example, you would have to switch from one to the other as your grammatical subject, just as you would when shifting between two ideas. In all cases, however, you'll find that it's easier to maintain unity if you follow the **SEES** pattern.

Avoid monotony

If most or all of the sentences in a paragraph have the same grammatical subject, how do you avoid sounding boring? Here are two ways:

1. **Use stand-in words.** Pronouns, either personal (*I, we, you, he, she, it, they*) or demonstrative (*this, that, these, those*), can stand in for the subject, as can synonyms (words or phrases that mean the same thing). The revised paragraph on artificial intelligence, for example, uses the pronouns *it* and *one* as well as the acronym *AI* to avoid too much repetition. Most well-written paragraphs have a liberal sprinkling of these stand-in words.

2. **"Bury" the subject by putting something in front of it.** When the subject is placed in the middle of the sentence rather than at the beginning, it's less obvious to the reader. If you take another look at the revised paragraph, you'll see that in a couple of sentences there is a phrase or two in front of the subject. Even a single word, such as *first*, *then*, *lately*, or *moreover*, will do the trick.

Link your ideas

To create coherent paragraphs, you need to link your ideas clearly. Linking words are those connectors—conjunctions and conjunctive adverbs—that show the relations between one sentence, or part of a sentence, and another; they're also known as "transition words," because they form a bridge from one thought to another.

Make a habit of using linking words when you shift from one grammatical subject or idea to the next, whether the shift occurs within a single paragraph or as you move from one paragraph to another. The following are some of the most common connectors and the logical relations they indicate:

Linking Word	*Logical Relation*
also	
and	
as well	
furthermore	
in addition	addition to previous idea
likewise	
moreover	
similarly	
alternatively	
although	
but	
by contrast	
despite, in spite of	
even so	change from previous idea
however	
nevertheless	
on the other hand	
rather	
yet	

accordingly

as a result

consequently

hence   } summary or conclusion

for this reason

so

therefore

thus

Enumerators such as *first*, *second*, *third*, *next*, and *finally* also work well as links.

Vary the length, but avoid extremes

Ideally, academic writing has a balance of long and short paragraphs. However, it's best to avoid the extremes—especially the one-sentence paragraph, which can only state an idea without explaining or developing it. A series of very short paragraphs is usually a sign that you have not developed your ideas in enough detail, or that you have started new paragraphs unnecessarily. On the other hand, a succession of long paragraphs can be tiring and difficult to read. In deciding when to start a new paragraph, remember always to consider what is clearest and most helpful for the reader. Because a paragraph balances general ideas with specific support, break extended paragraphs into smaller units. Adapt **SEES** to create short paragraphs including only a statement followed by elaboration.

Write a convincing conclusion

Endings can be painful—sometimes for the reader as much as for the writer. Too often, the feeling that one ought to say something profound and memorable produces a pretentious or affected ending. You know the sort of thing:

> The field of broadband multimedia networking is burgeoning. In the years to come, the communications system is destined for dramatic changes thanks to the monumental enhancements provided by electrical engineers.

This ending is too packed with clichés to be meaningful. The writer only seems relieved to get to the end.

Experienced editors say that many articles and essays would be better without their final paragraphs: in other words, when you have finished saying what you have to say, the best thing to do is stop. This advice may work for short essays, where you need to keep the central point firmly in the foreground and where you don't need to remind the reader of it. However, for longer pieces, where you have developed a number of ideas or a complex line of argument, you need to provide a sense of closure. Readers welcome an ending that helps to tie the ideas together; they don't like to feel as though they've been left dangling. And since the final impression is often the most lasting, it's in your interest to finish strongly. The following are two of the most convenient options.

The inverse funnel

The simplest conclusion is one that restates the thesis (**T**) in different words and discusses its implications (**I**). Sheridan Baker calls this the inverse funnel to contrast it with the funnel in the opening paragraph [1]. The discussion of the needs analysis, cited earlier for its funnel opening (p. 86), concludes with an "inverse funnel":

> (**T**) Having relied on a needs analysis to develop a problem statement, the designer is now ready to develop alternatives as solutions. (**I**) At the same time, he or she must appreciate the uncertain and iterative nature of the process and be prepared to rework, adjust, or even radically change the approach if necessary.

One danger in moving to a wider perspective is that you may try to embrace too much. When a conclusion expands too far it tends to lose focus. It's always better to discuss specific implications than to trail off into generalities in an attempt to sound profound.

The full circle

If you began your paper by posing a question or citing a startling fact, you can complete the circle by referring to it again in your conclusion, relating it to some of the insights revealed in the main body of your essay. This technique provides a solid sense of closure for the reader. Take advantage of the connectors listed on p. 91 to help.

Edit carefully

Editing doesn't mean simply checking your work for errors in grammar or spelling. It means looking at the piece as a whole to see if the ideas are well organized, well documented, and well expressed. It may mean making changes to the structure of the essay by adding some paragraphs or sentences, deleting others, and moving others around. Experienced writers may be able to check several aspects of their work at the same time, but if you are inexperienced or in doubt about your writing, it's best to look at the organization of the ideas before you tackle sentence structure, diction, style, and documentation. Follow the flowchart on p. 14.

Below is a checklist of questions to ask yourself as you begin editing. Far from all-inclusive, it focuses on the first step: examining the organization of your work. Since you probably won't want to check through your work separately for each question, you can group some together and overlook others, depending on your own strengths and weaknesses as a writer.

- Is my title concise and informative?
- Are the purpose and approach of this essay evident from the beginning?
- Are all sections of the paper relevant to the topic?
- Is the organization logical?
- Are the ideas sufficiently developed? Is there enough evidence, explanation, and illustration?
- Would an educated person who hasn't read the primary material understand everything I'm saying? Should I clarify some parts or add any explanatory material?
- In presenting my argument, do I take into account opposing arguments or evidence?
- Do my paragraph divisions make my ideas more coherent? Have I used them to keep similar ideas together and signal movement from one idea to another?
- Do any parts of the essay seem disjointed? Should I add more transitional words or logical indicators to make the sequence of ideas easier to follow?
- Do my conclusions accurately reflect my argument in the body of the work?

An additional approach is to devise your own checklist based on comments you have received on previous assignments. This is particularly useful when you move from an overview of organization to the close focus on sentence structure, diction, punctuation, spelling, and style. If you have a particular weak area (for example, wordiness or run-on sentences), be sure give it special attention. Keeping a personal checklist will save you from repeating the same old mistakes.

Writing Commentary on a Book or Article

The assignment to evaluate a book or article usually calls for more than a summary of contents, but less than the sophisticated literary essay you might expect from the *Globe and Mail*. If you are required to prepare something for a course, it will likely be an analytic report containing some evaluation. The following guidelines should help.

Approach an evaluative report the same way you would approach an essay, by writing an outline. Begin with an introduction; then follow with a summary and an evaluation. Publication information is usually listed at the beginning but can also be placed at the end.

Prepare an introduction

In your introduction, provide all the background information necessary for a reader who is not familiar with the work. Here are some of the questions you might consider:

- What is the work about? Is the title pertinent and useful as a guide to the contents?
- What is the author's purpose? What kind of audience is he or she writing for? How is the topic limited? Is the central theme or argument stated or only implied?
- How does the work relate to others in the same field of interest?
- What are the author's background and reputation? What books or articles has he or she written?
- Are there any special circumstances connected with the writing? For example, was it written with the co-operation of particular scholars or institutions? Does the subject have special significance for the author?
- What kind of evidence does the author present to support his or her ideas? Is it reliable and up to date?

Not all of these questions will apply to every work, but an introduction that answers some of them will put your reader in a much better position to appreciate what you have to say in your evaluation.

Include a summary

You cannot analyze anything without discussing its contents. You may choose to present a condensed version of the contents as a separate section, to be followed by your evaluation, or you may prefer to integrate the two, assessing the author's arguments as you present them.

Focus on the evaluation

In evaluating the work, you want to consider some of the following questions:

- How is the work organized? Does the author focus too much on some areas and too little on others? Has anything been left out?
- How has the author divided the work? Are the divisions valid? Do titles and subtitles accurately reflect each section's contents?
- What kind of assumptions does the author make in presenting the material? Are they stated or implied? Are they valid?
- Does the author accomplish what he or she sets out to do? Does the author's position change in the course of the writing? Are there any contradictions or weak spots in the arguments? Does the author recognize those weaknesses or omissions?
- What documentation does the author provide to support the central theme or argument? Is it reliable and current? Is any of the evidence distorted or misinterpreted? Could the same evidence be used to support a different case? Does the author leave out any important evidence that might weaken his or her case? Is the author's position convincing?
- Does the author agree or disagree with other writers who have dealt with the same material or problem? In what respects?
- Is the work clearly written and interesting to read? Is the writing repetitious? Too detailed? Not detailed enough? Is the style appropriate? Or is it plodding and full of jargon? Is it flippant?
- Does the work raise issues that need further exploration? Does it present any challenges or leave unfinished business for the author or other scholars to pick up?
- If you are evaluating a book with an index, how good is this index?
- Are there illustrations or graphics? Are they helpful?
- To what extent would you recommend this work? What effect has it had on you?

Remember that your job is not to interpret the contents but to highlight strengths and weaknesses. In short, be objective and fair.

Chapter Checklist

Whether you're writing a reflection paper, an essay, or a book review, be prepared to complete multiple drafts attending to the following stages.

- ☐ Make sure you understand the assignment. Narrow the topic to one which suits the required length, purpose, and audience.
- ☐ Develop your thesis, and outline your development.
- ☐ Complete your research as well as all required readings
- ☐ Write a first draft, modelling your paragraphs upon established patterns.
- ☐ Take time to revise your draft as necessary to develop your ideas further and add coherence.
- ☐ Edit and proofread carefully before submitting.

Notes

1. S. Baker, *The Longman Practical Stylist*. White Plains, NY: Longman, 2005, p. 80.
2. J.F. Trimmer, *Writing with a Purpose*, 14th ed. Boston: Houghton Mifflin, 2005, pp. 62–63.

Documentation

9

> **Chapter Objectives**
> - Incorporating quotations in your writing
> - Documenting your sources using IEEE, APA, CSE, Chicago, and MLA styles

Introduction

Much of the writing you do requires you to consult secondary sources for project ideas as well as to keep up with current research. Bibliographic software like RefWorks, BibTex, EndNote, and Zotero makes it easy for you to keep track of information on your sources and also to format both notes and reference lists. Be sure to check what's available through your school's library.

Although scientific writing calls for few direct quotations, it is still essential to acknowledge your sources, not just if you quote directly from them but also when you refer to observations, conclusions, theories, or ideas presented in them. If you don't acknowledge these sources, you let your reader assume that the words, ideas, or concepts are yours. Such an omission is considered plagiarism, and penalties are severe (see pp. 26–30).

Of course, the purpose of documentation is less to avoid charges of plagiarism than to situate your work within the body of knowledge in your discipline. Academic writing is anchored on the premise that researchers are indebted both to scholars who came before them and to colleagues. By documenting your sources, you show that you recognize your indebtedness and are ready to make your own contribution to your field.

Using Direct Quotations

If you are not writing a technical paper, judicious use of direct quotations can add authority to your writing. However, never quote a passage just because it sounds impressive; be sure that it really adds to the discussion. Perhaps it expresses an idea with special force or cogency or gives substance to a debatable point. Quote directly only when you need to present the source's exact words. (If you allude to a source, rather than quote directly, follow the conventional methods of referencing presented later in this chapter.) The following are situations that justify word-for-word quotations from primary or secondary sources:

- **Literary content.** An analysis or evaluation of someone's writings (prose or poetry) calls for exact quotations of the literary text.
- **Expert opinion.** A representative statement or a memorable comment from a recognized authority is worth citing, especially if it is the subject for discussion.
- **First-hand reports.** In media accounts and news releases, an accurate record from witnesses adds necessary specificity.

When you are convinced that only a direct quotation will serve your purposes, the following are guidelines you should follow:

1. Integrate the quotation so that it makes sense in the context of your discussion and fits grammatically into your sentence:

    ✗ Bill Gates could not foresee the future. "640K ought to be enough for anybody" is highly ironic.

    ✓ Bill Gates' 1981 prediction that "640K ought to be enough for anybody" is ironic in light of exponential leaps in computer engineering.

2. If the quotation is fewer than four lines long, include it as part of your text, enclosed in quotation marks. If the quotation is four lines or longer, set it as a block of free-standing text, double-spaced and indented from the left-hand margin (without quotation marks). If the quotation consists of more than one paragraph, indent the first line of the second and all subsequent paragraphs an additional three spaces.

    If you are quoting brief lines of poetry, they can be included as part of your text. Use a slash (/) to indicate the end of a line. For verse quotations longer than four lines, write the words line for line as originally written.

3. Be accurate whenever you are using quotation marks. Reproduce the exact wording, punctuation, and spelling of the original. (You can always acknowledge a typo or mistake in the original by inserting the

word *sic* in square brackets after it—see pp. 215–216). If you want to insert an explanatory comment of your own into a quotation, enclose it in square brackets:

> As Jones points out, "Biology has a statistical machine [cladistics] to order the world."

4. If you want to omit something from the original, use ellipsis points. (See p. 224 for details.)

Documenting Your Sources

Your purpose for including a list of references at the end of your document is to make it as easy as possible for any reader to track your sources. That's why each reference includes as much of the following information as possible:

- name of author or authors, in the order listed in the original work
- publication date
- title of article, posting, paper, or chapter
- title of book, journal, periodical, or collection, including edition if applicable
- name of editor or translator, identified by the abbreviations "ed." or "trans."
- place of publication and publisher, for books only
- volume number, issue number, and page numbers for journals and periodicals
- site sponsor/organization for websites
- DOI (digital object identifier) or URL for electronic sources

List only those works that you have actually referred to in the text of your writing. As well, it's important to cite the exact source you consulted. Don't list a print version if your source was electronic.

There are many systems of documentation, and the one you use depends on your subject as well as on the preference of your instructor, your department, or your employer. It makes sense from the beginning to find out whether there is a preferred documentation style and set of guidelines for its use. For easy reference, the most common systems are presented later in this chapter. Remember, though, that style guides are constantly undergoing revision, especially with the wealth of electronic information available. Always check the website or the latest edition of the relevant manual to be sure that you have the most up-to-date information.

Be aware as well that technical societies may produce their own specifications, spelled out on their websites. If you are preparing a paper for publication, you are expected to follow the guidelines of the organization exactly, and there may be major or minor variations from the standards you follow in other contexts. It is a mark of professionalism to keep to requirements exactly.

There is one absolute principle in documenting anything—the demand for consistency. Once you have committed to following a particular referencing style, you must continue to use it throughout whatever it is that you are writing.

Even if you have not been told to follow a specific style manual or set of guidelines, you are still expected to acknowledge your sources. In scientific writing, the most common system of documentation is also the simplest and most economical. The following guidelines outline the system for IEEE publications [1], which we use in this handbook. Keep to these conventions if no others have been specified for you.

IEEE Style

In-text citation sequencing

Number your citations consecutively, putting the reference numbers in square brackets and punctuating afterwards if necessary. The first reference in your text will be [1], the next new reference will be [2], and so on. Once a source has been assigned a number, it is referred to by that number whenever it appears in the text:

> Labossière's groundbreaking study [1] was first challenged by Gormon [2] and later disputed by Huang [3]. Gormon's later work [4] confirmed the validity of Labossière's original hypothesis [1].

Putting reference numbers in square brackets both in the text and in the list of references is a preferred alternative to raised numerals that might be confused with superscript in scientific abbreviations or exponents. If you are referencing several sources at once, include all reference numbers in the same citation, giving each its own set of square brackets. Refer to multiple references this way:

> Ultrasound images contain speckle noise [1]–[3] that makes tumours difficult to detect by eye alone.

If you make a later reference to a work you have already cited, use the number you assigned the work originally. List the work only once by number in your reference list.

Reference lists

The IEEE regularly refers writers to the *Chicago Manual of Style* (see p. 113) for questions of style, but the reference guidelines outlined in the *IEEE Editorial Style Manual* [1] provide formatting advice. This method for preparing a comprehensive list of sources at the end of your work merges traditional formats for endnotes and bibliographies. If you are using the in-text citation method, you list references in numerical order, according to their order of appearance in your work. Put the bracketed number flush with the left margin, and use a hanging indent if your entry extends beyond the initial line. To conserve space, use initials rather than first names for authors, and use standard abbreviations for the names of journals, conferences, conference proceedings, and organizations. If you are referring to a page or two, rather than the entire work, list only the page numbers. Follow the punctuation conventions below:

[1] I. Author and U. Writer, "Title of chapter," in *Book*, T. Smith, Ed. City, State, Country: Publisher, Year, pp. xx-xx.

[2] I. Author and U. Writer, "Title of article," *Abbrev. Title of Periodical, vol. x*, no. x, pp. xx–xx, Abbrev. Month, Year.

[3] I.M. Author. (year, month, day). Title. *Source* [Medium]. *volume* (issue), pp. xx–xx. Available: URL or DOI.

The models above italicize titles and volume numbers, but use regular type for the titles of chapters or articles. Always check with your instructor or a librarian if you are in doubt.

The following examples provide IEEE models for scientific references including titles in italics. Remember to give as much information as necessary for your reader to locate each source.

Book with one author

The author's name follows the citation number and is not inverted. Capitalize important words in the title:

[1] R. Van Meter, *Quantum Networking*. Hoboken, NJ, USA: John Wiley & Sons, 2014.

Book with more than one author
If the book you are citing has two or more authors, list all the names, separated by "and" (and commas, in the case of three or more authors).

> [2] J.P. Goedbloed, R. Keppens, and S. Poedts, *Advanced Magneto-hydrodynamics.* New York, NY, USA: Cambridge University Press, 2010.

Book with a group or corporate author
> [3] University of Chicago Press, *The Chicago Manual of Style*, 16th ed. Chicago, IL, USA: University of Chicago Press, 2010.

In the case of a revised or subsequent edition, include this information following the title, as shown above. (A revised edition would be shown as "rev. ed.")

Book with an editor, compiler, or translator
If the book has an editor or compiler and no author, give the editor's or compiler's name followed by "Ed." or "Comp.":

> [4] V. I. Kodolov, Ed., *Nanostructures, Nanomaterials, and Nanotechnologies to Nanoindustry.* Toronto, Canada: Apple Academic Press, 2015.

If the book has a translator, editor, or compiler as well as an author, put the author's name before the title and the other name after the title, introduced by the appropriate abbreviation.

> [5] N. Copernicus, *On the Revolutions of the Heavenly Spheres.* Trans. A. M. Duncan. New York, NY, USA: Barnes & Noble, 1976.

Selection in an edited work
The title of the selection is placed in quotation marks and in lower case but for the first letter of the first word of the title and subtitle. It follows the author's name and precedes the name of the edited work. Notice that the word *in* precedes the name of an edited book but is *not* used before the names of journals, periodicals, magazines, or newspapers containing a referenced article:

> [6] R.F. Miller, "Geotourism, mining and tourism development in the Bay of Fundy," in *Mining Heritage and Tourism: A Global Synthesis,* M.V. Conlin and L. Jolliffe, Eds. New York, NY, USA: Routledge, 2011. pp. 211–225.

Article in a journal or periodical

> [7] J. Shapiro, "Electromagnetic compatibility for system engineers," *IETE Technical Review*, vol. 28, no. 3, pp.70–77, May–June 2011. doi: 10.4103/0256-4602.74505

All electronic articles have a DOI (digital object identifier), which offers a permanent link to the publication. To find a missing DOI, use http://www. crossref.org.

If the work has not yet appeared in print, put "to be published" if the work has been accepted for publication. Use "submitted for publication" if it has not yet been accepted:

> [8] M. Foumani, A. Khajepour, and M. Durali, "Optimization of engine mount characteristics using experimental numerical analysis," *J. Vibration and Control,* to be published.

Article in a newspaper

If the section of the newspaper containing the article you are referencing is identified, give its name, number, or letter:

> [9] J. Ball, "Wind power hits a trough," *Wall Street Journal,* p. A20, April 6, 2011.

If the article is unsigned, begin your reference with the title of the article.

Technical report

Even if a technical report has been printed in house (i.e., by the organization itself), you list it as a publication. Include all necessary information for locating it, including the report number abbreviated as "Rep."

> [10] "A summary of NRC construction housing activities for 2012: A report prepared for the Canadian Home Builders' Association," NRC Institute for Research in Construction, NRC, Rep. 68187, Feb. 2013.

Paper presented at a conference

Cite this as you would a journal article, but include the words "presented at" to indicate that you are referencing a paper you attended:

[11] B. Richter, "Chasing water: The imperative to move from scarcity to sustainability," presented at the 10th Intl. Symp. on Ecohydraulics, Trondheim, Norway, June 26, 2014.

Otherwise, use the conventional reference format for a publication.

Dissertation
Refer to an unpublished dissertation this way:

[12] V. Bevan, "Sediment budget of an urban creek in Toronto." PhD dissertation. Dept. Civ. Eng, Univ. Waterloo, Waterloo, Canada 2014.

Lecture notes
Give as much detail as necessary for the reader to locate the material. For material included in courseware packs, give original publication information as well.

[13] M.P. Anderson and W.W. Woessner, "Profile models," in *Applied Groundwater Modeling*, 2nd ed. Elsevier: Academic Press, 2011. pp. 172–194. Included in C. MacGregor, course notes. SD 347, Conestoga College, Waterloo, Canada, Fall 2014.

For references to material (a slide, for example, or notes on the board) coming from a class you attended, acknowledge the source by providing the details, the course name, institutional information, and date.

[14] C.R. Ellis, "Pavement stress graphs: Highway 404, 2011," CIVL 460, Queen's Univ., ON, Nov. 23, 2014.

Patent
Include the inventor's name, the title of the invention, the patent number, and date of issue:

[15] M. Bureau, "Fluid Recovery," Can. Patent CA 2610221, Sept. 2, 2014.

Multimedia sources
Specify the type in square brackets:

[16] D. Suzuki, Chris Hatfield: The Man who Tweeted Earth [DVD], 44 min. CBC, 2013.

Website

The most up-to-date information is available instantly, but it can change. Especially if you are citing a wiki, you should identify the date you accessed the site (to show when your information was current). List this information at the end of your citation. References must include enough information about the material, the site, and the sponsoring organization that the reader can access the source. For all online sources, then, provide the full DOI if available; otherwise, provide the URL, not just the home-page address. Be sure to report the DOI or the URL accurately.

[17] S. Arbesman. (2014, Oct. 2). From 53 kilobytes of RAM in 1953. [Online]. Available: http://www.wired.com/2014/10/53-kilobytes-ram-1953/

Note that if you have to break a URL so that it fits properly on a line, do so after a slash. Don't add extra spaces or use hyphens.

Alphabetical Referencing

Some professors will ask you to list your references in alphabetical order at the end of your document rather than in their order of appearance in the text. Follow their recommendations by assigning a number to each reference only once everything has been alphabetized. (For entries, use the same format as that described above, but reversing first names and surnames.) Thus your reference list will be in numerical and alphabetical order at the same time. To complete the process, go back to your document and make sure your references are keyed to the numbers listed at the end. When the reference list is alphabetized, numbers in the text will therefore not appear in sequence:

Ultrasound images contain speckle noise [2],[5],[6] that makes tumours difficult to detect by eye alone.

Standard Style Sheets

If you have professors, colleagues, or situations calling for a specific reference style, it is likely to be one of APA, CSE, Chicago, or MLA. Below you will find examples of these formats as well as additional references for you to consult. There are minor differences distinguishing each of these, so be sure to follow the conventions exactly. If you have access to RefWorks or other bibliographic software, simply click on your chosen style sheet to have your references formatted automatically.

APA Style

The American Psychological Association system of documentation is the one most commonly used in the social sciences, business, and nursing. The APA system uses in-text citations, which name authors and dates in parentheses after the information cited, rather than providing reference or note numbers. Complete bibliographical information is included in a list of references at the end. Some examples are listed below. For more detailed information, consult the *Publication Manual of the American Psychological Association* (6th ed.). The APA website at http://apastyle.apa.org/ also provides current information on APA Style.

In-text citations
Source with one author
If the author's name is given in the text, cite only the year of publication in parentheses.

> Helmsteadt (2014) made a significant contribution to the climate change debate.

Otherwise, give both the name and the year:

> The latest analysis (Christensen, 2015) has disproved government statements.

Source with more than one author
If the work you are citing has two authors, include both names every time you cite the reference in the text. APA uses an ampersand (&) when the names are in parentheses, but uses "and" in the text:

> Earlier research showed that it was possible to redesign agro ecosystems for environmental sustainability (Hill & Henning, 2013). Todd and Voisin (2014) later showed that rural communities could afford to implement such systems only with heavy government subsidies.

If there are three, four, or five authors, cite all the names when the reference first occurs and afterwards cite only the first author, followed by "et al.":

> Reid, Jensen, Nikel, and Simovska (2015) offered new insights on sustainability projects. [. . . .] Research showed that public school students were capable innovators (Reid et al., 2015).

If the work you are citing has six or more authors, cite only the surname of the first author followed by "et al."

Source with a group or corporate author

Corporations, associations, and government agencies serving as authors are usually given in full each time they appear. Some group authors may be named in full the first time and abbreviated in subsequent citations if this provides the reader with enough information to easily locate the entry in the reference list:

> A recent study examines the effectiveness of treating low-back pain in the first two weeks following injury (Institute for Work & Health [IWH], 2014). . . . The same study also looks at the effectiveness of bed rest in speeding up recovery (IWH, 2014).

Work with no known or declared author

If the work you are referencing has no known or declared author, cite the first few words of the title, as well as the year:

> In the pyroelectric sensor, the infrared light generates surface electric charge off a substrate ("Motion Sensing Lights," 2015).

[The full title of the source is "Motion Sensing Lights and Burglar Systems Work, Study Shows."]

Specific parts of a source

If you are referring to a particular part of a source, indicate the page, chapter, figure, table, or equation. Always give page numbers for quotations:

> (Felderhaus, 2012, p. 299)
> (Christenssen & Schor, 2014, pp. 2601–2602)
> (Stenson, 2013, Chapter 2)

Note that the APA abbreviates page numbers with "p." for a single page and "pp." for several pages.

Online sources
In-text citations for electronic sources use the same formatting principles outlined above for print sources, with the following exceptions:

- If your source has no page numbers, use the paragraph number, if one is available, preceded by the abbreviation "para.":

  (Edson, 2015, para. 10)

- If sections, pages, and paragraphs are not numbered, cite the heading and the number of the paragraph following it to direct the reader to the specific location you are referring to:

  (Black, 2014, Introduction, para. 6)

References
The following conventions are used for recording entries in an APA *References* section:

- Entries begin with the author's surname, followed by his or her initials (full given names are not used).
- In entries for works with more than one author, all authors' names are reversed, with the name of the last author preceded by an ampersand (&) rather than "and."
- The date of publication appears immediately in parentheses after the names of the author(s).
- Entries for different works by the same author are listed chronologically. Two or more works by the same author with the same publication year are arranged alphabetically by title.
- For titles of books and articles, only proper nouns and the first word of the title and of the subtitle (if there is one) are capitalized. However, titles of journals are capitalized as usual.
- Include the DOI (digital object identifier) for articles from databases and e-journals.
- Titles of articles or selections in books are *not* enclosed in quotation marks. Book and journal titles are italicized.

For references to electronic sources, no retrieval date is needed. Provide the URL with "Retrieved from" if you are providing a web address that is freely accessible or "Available from" if it comes from a source that must be purchased (a database or subscription service, for example).

Book with one author
Lane, C.A. (2013). *On fracking*. Victoria BC: Rocky Mountain Books.

Book with more than one author
Bell, M. & Buckley, C. (2010). *Solid states: Concrete in transition*. New York: Princeton Architectural Press.

For books with eight or more authors, list only the first six followed by three ellipses and the last author's name.

Book with a group or corporate author
European Commission Directorate-General for Energy. (2011). *Renewables make the difference*. Luxembourg: Publications Office of the European Union.

Book with an editor or editors
Weiss, N. & Koschak, A. (Eds.). (2014). *Pathologies of calcium channels*. Heidelberg DE: Springer.

Selection in an edited book
Galison, P.L. (2011). Tools and innovation. In R.Y. Chiao, (Ed.), *Visions of discovery: New light on physics, cosmology, and consciousness* (pp. 23–30). New York, NY: Cambridge University Press.

Note that the page numbers of the selection are given, preceded by "pp."

Article in a journal or magazine
Arjoranta, J. (2014). Game definitions: A Wittgensteinian approach. *Game Studies, 14* (1). Retrieved from http://gamestudies.org/1401/articles/arjoranta

If there is a volume number, the page range is given without "pp." If there is no volume number, include "pp." to indicate that the numbers refer to pagination.

When a journal has continuous pagination, the issue number should not be included. If each issue begins on page 1, give the issue number in parentheses before the page numbers and immediately following the volume number, with no punctuation separating them and a comma following the parentheses. The volume number is italicized or underlined; the issue number and its parentheses are not. Include the digital object identifier (DOI), if one is provided; otherwise, provide the URL:

Osoba, L.O., Sidhu, R.K., & Ojo, O.A. (2011). On preventing HAZ cracking in laser welded DS Rene 80 superalloy. *Materials Science and Technology, 27*(5), 897–902. doi:10.1179/026708309X12560332736593

For monthly or bimonthly magazines, give the month(s) in full; for weekly magazines, give the month (in full) and day.

Multimedia sources
Be sure to include either the DOI or URL for electronic sources.

Walters, R. (2013, April 27). Reusing spray foam cans. [YouTube]. *Ron's Stuff.* Retrieved from http://www.youtube.com/watch?v=PVOfoUD5Mvl

Website
Environment Canada. (2014 Oct. 19). Air quality health index. Retrieved from http://weather.gc.ca/airquality/pages/onaq-004_e.html

Remember to break a URL that goes to another line after a slash or before a period. Do not insert (or allow your word processing program to insert) a hyphen at the break.

Message posted to a newsgroup or mailing list
Give as much information as possible to allow a reader to track the source.

Bryant, M.D. (2015 February 4). DNAPL contamination [Electronic mailing list message]. Retrieved from news://sci.aptenv.poll

Because e-mail messages and tweets are not recoverable, do not list such communications in the *References* section, but do reference them in the text of your document.

CSE Style

The Council of Science Editors recommends a citation-sequence system using raised numerals to direct the reader to a list of references at the end of the paper, like IEEE. This system is commonly used in the medical sciences. The following guidelines are based on the principles laid out in the eighth edition of *Scientific Style and Format: The CSE Manual for Authors, Editors, and Publishers* (2014). (Details about other CSE styles can be found in this latest edition.)

- Superscript numbers in the text correspond to numbered references in the *References* section at the end of the document.
- If several sources are being referenced at once, all relevant reference numbers should be given in the same citation, separated with commas and no spaces. With a sequence of three or more citation numbers (e.g., 7, 8, 9), use only the first and last number in the sequence, separated with a hyphen:

  Several studies[1,5,12–15] have shown . . . .

- Include only material you have read and referred to in the text of your paper. If you want to point your reader to sources you have consulted but not referred to in your document, include them in a section labeled *Supplemental References* or *Additional Sources*.
- In the *References* section, list references in order of their appearance in the text. For each entry, follow these conventions exactly:
  1. The note number followed by a period and two spaces.
  2. The author's last name followed by initials without spacing or punctuation and then a period. For multiple authors, list the names up to a maximum of ten, separating the names themselves with commas. Note that "and" is not used. End the list with a period.
  3. The title of the article or selection, followed by the title of the journal or book, with *no* underlining or italics or quotation marks. Only the first word of the title and proper nouns are capitalized. For journals, abbreviate titles and capitalize major words.
  4. The place of publication followed by a colon.
  5. The abbreviated name of the publisher followed by a semicolon.
  6. The date of publication followed by a period.
  7. The range or total number of pages. For books, put the total number of pages followed by "p." For articles, give the page range with the fewest possible digits. With journals, page references appear after a colon.
  8. Put additional non-publishing information (e.g., type of media, citation dates) in square brackets.
  9. Include a DOI (digital object identifier) if available; otherwise, provide a URL.

References

Book

1. Kluyver CA, Pearce JA. Strategy: a view from the top. Englewood Cliffs (NJ): P-H; 2012. 304 p.

Note that the last element of the entry ("304 p.") indicates the total number of pages in the book. You can find the extent of any library material in its library catalogue citation.

Contribution in an edited book

2. Heinzel A, König U. Nanotechnology for fuel cells. In: Leite ER, editor. Nanostructured materials for electrochemical energy production and storage. New York (NY): Springer; 2011. p. 151–84.

Article in a journal

3. Vassilicos JC. Dissipation in turbulent flows. Annu Rev Fluid Mech 2015; 47(1): 95–114.

Note that the CSE abbreviates names of journals and magazines without adding periods.

Article in an online journal

4. Kurtuldu G, Jessner P, Rappaz M. Peritectic reaction on the Al-rich side of Al–Cr system. J All Com. 2015 [cited 2015 Mar 30]; 621(1): 283–286. http://www.sciencedirect.com/science/article/pii/S0925838814023470. doi: 10.1016/ j.jallcom.2014.09.174

Always include a DOI (digital object identifier) if available (see p. 103).

Website

5. Masse B. Biologics program. Ottawa: National Research Council Canada; 2014 [modified 2014 Feb 24; accessed 2015 Jan 3]. http://www.nrc-cnrc.gc.ca/eng/solutions/collaborative/biologics_index.html

Because web information can change regularly, always identify the date of the most recent version.

Multimedia Source

6. Canadarm: what is this big arm? [video]. CBC Digital Archives. 1979 Aug 30, 10:32 minutes. [accessed 2015 May 17]. http://www.cbc.ca/player/Digital+Archives/Science+and+Technology/ID/1739503562/

Chicago Manual of Style

The Chicago Manual of Style (CMS) outlines two methods of documentation. The *author–date* system, preferred in the natural and social sciences, follows the same principles as APA style with only minor stylistic differences. The *notes-and-bibliography* method uses superscript numerals to direct the reader to footnotes at the bottom of the page or endnotes listed separately at the end of the document. In scientific writing, footnotes or endnotes are used not to cite references but to expand upon the content of the paper—on the very rare occasions where such elaboration is impossible in the text. Some journals, for example, will use a footnote to provide biographical details of the author(s). In most of your work, you can include the reference in the text itself by name, year, and page in parentheses like so: (Northey and Jewinski 2016, 34).

Despite the convenience of footnoting software, use notes *only* if you are specifically directed to do so. The following are some examples of notes using the Chicago notes-and-bibliography method. For additional examples, consult *The Chicago Manual of Style: The Essential Guide for Writers, Editors, and Publishers* (16th ed.) or the University of Chicago Press/Questions & Answers website, available at http://www.chicagomanualofstyle.org/qanda/latest.html

Notes

Footnotes appear at the bottom of the page or column that contains the reference. Endnotes are placed on a *Notes* page after any appendices and before a bibliography. All notes are single-spaced and typed with an indented first-line (as in the examples below). Although the note numbers are superscript where they appear in the text, the note numbers that precede the notes themselves are not. If you have been asked to cite your references with CMS notes, prepare them to meet the following conventions.

Book with one author

The name of the author follows the note number and is not inverted. It is followed by a comma and the italicized title of the book, with all important words capitalized. The publication details follow the title in parentheses, and the page numbers are added without an abbreviated "p.":

1. Dan Marinescu, *Classical and Quantum Information* (Maryland Heights, MO: Academic Press, 2012), 87.

Book with more than one author

If the book you are referencing has two or three authors, all of the authors' names are given, separated by "and" and (in the case of three or more authors) commas:

> 2. James Ambrose and Patrick Tripeny, *Simplified Engineering for Architects and Builders,* 11th ed. (Hoboken, NJ: Wiley, 2011), 155.

If the book you are referencing has more than three authors, you can use the name of only the first author, followed by either "et al." or "and others" and a period, without intervening punctuation.

Book with a group or corporate author

> 3. University of Chicago Press, *The Chicago Manual of Style,* 16th ed. (Chicago: University of Chicago Press, 2010), 244.

In the case of a revised or subsequent edition, include this information following the title, as shown above. A revised edition would be "rev. ed."

Book with an editor, compiler, or translator

If the book you are referencing has an editor or compiler and no author, give the editor's or compiler's name first, followed by "ed." or "comp.":

> 4. Kent Pinkerton and William N. Rom, eds., *Global Climate Change and Public Health.* (New York: Humana Press, 2014), 48.

If the book you are referencing has a translator, editor, or compiler as well as an author, the author's name should come before the title with the translator's, editor's, or compiler's name following the title, introduced by the appropriate abbreviation:

> 5. Madeleine Ferrier, *Sacred Cow, Mad Cow: A History of Food Fears,* trans. Jody Gladding (New York: Columbia University Press, 2006), 24–31.

Selection in an edited book

The title of the selection, set in quotation marks, follows the author's name and precedes the name of the edited work. Note that a title within a title is enclosed in quotation marks:

> 6. Elizabeth A. Gibson and Anders Hagfeldt, "Solar Energy Materials," in *Energy Materials,* ed. Duncan W. Bruce, Dermot O'Hare, and Richard I. Walton (Chichester, West Sussex: Wiley, 2011), 246–278.

Article in a journal

If an issue number is given, it follows the volume number, separated by a comma and "no." In a reference to the article as a whole, the entire page range should be included. A reference to a particular section should give relevant page numbers only. Include the DOI (digital object indicator) if the article comes from an online journal:

7. Ron Zevenhoven, Martin Falt, and Luis Pedro Gomes,"Thermal Radiation Heat Transfer: Including Wavelength Dependence into Modelling, *International Journal of Thermal Sciences* 86, (Dec, 2014): 189-197. doi: 10.1016/j.ijthermalsci.2014.07.003

Article in a magazine

Even if a magazine is numbered by volume and issue, it is usually cited by date only:

8. Charlie Gillis and Chris Sorensen, "Finding Franklin," *Maclean's*, Sept. 22, 2–14, 40–44.

Website

To cite a website or other electronic source, include as much of the following as can be determined: author of the content, title of the page, title or owner of the site, and URL or DOI.

9. "Aggressive Driving and Road Rage," Government of Ontario. http:// www.mto.gov.on.ca/english/safety/topics/aggressive.shtml.

Bibliography

The bibliography is a list of all the sources you have used in your paper, including works you may have consulted but not referred to directly. It is placed on a separate page at the end, arranged alphabetically by the authors' last names, single-spaced, with hanging indents. If you are asked to keep primary and secondary source materials separate, or list online sources separately from paper sources, use subheadings. Do not number your entries.

Book with one author

Funk, McKenzie. *Windfall: The Booming Business of Global Warming*. New York: Penguin, 2014.

Book with more than one author

If the book you are referencing has two authors, invert the name of the first author only and separate the names with a comma and "and":

> Tomecek, Stephen M., and Steve Tomecek, *Global Warming and Climate Change*. New York: Chelsea House Publications, 2011.

If the book you are citing has four to ten authors, list only the first author, followed by "et al." in your note, but your bibliography should list all authors. For works of more than ten authors, you may also follow the first author's name (inverted) with a comma and "et al." or "and others":

> Hausladen, Gerhard, Michael de Saldanha, Petra Liedl, Hermann Kaufmann, Gerd Hauser, Klaus Fitzner, Christian Bartenbach et al. *ClimateSkin: Building-skin Concepts that Can Do More with Less Energy*. Basel: Birkhauser, 2008.

Book with a group or corporate author

> United Nations Development Programme. *Global Environmental Outlook* 3. London: Earthscan, 2002.

Book with an editor or translator

If the book has an editor and no author, give the editor's name first, followed by the abbreviation "ed." or "eds." if there are multiple editors:

> Nicholson, Simon, and Paul Kevin Wapner, eds. *Global Environmental Politics: From Person to Planet*. Boulder, CO: Paradigm Press, 2015.

If the book has an editor or translator as well as an author, give the author's name first and give the translator's or editor's name after the title, introduced by "Edited by" or "Translated by":

> Ferrier, Madeleine. *Sacred Cow, Mad Cow: A History of Food Fears*. Translated by Jody Gladding. New York: Columbia University Press, 2006.

Selection in an edited book

> Spielgelman, Jonah. "Management to Industrial Ecology." In *Linking Industry and Ecology: A Question of Design*, edited by Raymond Côté, James Tansey, and Ann Dale, 225–241. Vancouver: UBC Press, 2006.

Article in a journal

Since a bibliographic entry is a reference to the article as a whole, the entire page range should be given:

> Mwafy, Aman. "Assessment of Seismic Design Response Factors of Concrete Wall Buildings." *Earthquake Engineering and Engineering Vibration* 10, no.1 (March 2011): 115–127. doi:10.1007/s11803-011-0051-7

If you are referencing an unsigned article, the entry should begin with the title of the article.

The preceding example lists an electronic journal, but the same guidelines apply to electronic magazines and newspapers: follow the format of the print counterparts, with the addition of the DOI or URL and, if the information is time-sensitive, the date of access. To list a website, include as much of the following as can be determined: the author of the content, the title of the page, the title or owner of the site, and the URL or DOI.

> Kukulka, David et al. "Thermal Conductivity of Natural Rubber Using Molecular Dynamics Simulation." *Journal of Nanoscience and Nanotechnology* 15, no. 4 (April 2015): 3244-3248. doi: 10.1166/jnn.2015.9640

Multimedia Source

> Laszewski, Nick. "DIY Battery Backup Sump Pump." *Lifestyle*. YouTube. Mar. 4, 2014. 10:15. http://www.youtube.com/watch?v=7twLj5MDCO4

Because electronic communications can be difficult or impossible to access, Chicago style recommends making only in-text references to such sources as e-mails or tweets. They are not normally mentioned in a bibliography although you still must cite them in a footnote or endnote.

MLA Style

If you are taking elective courses in the humanities, the Modern Language Association is the accepted authority for documenting your sources. MLA style uses in-text citations, which give the author's last name and the page number in parentheses after the information cited. Complete bibliographical information is then given in an alphabetical list, titled *Works Cited*, on a separate page at the end of the paper.

The following examples illustrate the most common types of references as they would appear in citations and in the *Works Cited* list. If you don't see what you're looking for, consult the *MLA Handbook for Writers of Research Papers* (7th ed.).

In-text citations

Book or article with one author

Put in parentheses (round brackets) only the information needed to identify a source in your *Works Cited*—usually the author's (or editor's) last name and the page number of the text referred to:

> Einstein later described his original theory of relativity as "child's play" (Lawden 46).

Place the parenthetical reference where a pause occurs, usually at the end of the sentence or clause it documents. Note that there is no punctuation between the author's name and the page number.

If the author's name is already given in the text, put in parentheses only the page number(s):

> Harris sees the logic in Hegel's argument (72–74).

If you are citing an entire work, try to include the author's (or editor's) name in the text rather than in parentheses:

> In *Elements of Relativity Theory,* Lawden considers both historical and scientific contexts of Einstein's work.

Source with more than one author

If the work has two or three authors, include all of the names in the citation:

> The same argument was applied to the universities (Matthews and Steele 50–62).

If the work has four or more authors, either use only the first author's last name and "et al." or give all the last names; use the same form in the *Works Cited* section:

> The result, some claim, is "cultural suicide" (Jones et al. 42).

Book or article with a group or corporate author

If you are citing a document prepared by a corporate author or government agency, give the name of the organization or agency, omitting any article (such as "the" or "a") within your parentheses. Abbreviations are permitted for government agencies but not for corporations. It is often clearer if you include this information in your sentence rather than in your citation:

> Until a correction appeared on the website, the Canada Post Corporation showed the abbreviation for both Nunavut and Northwest Territories as "NT" (14).

Two or more works by the same author

If you need to refer to more than one work written by the same author, use a shortened version of the appropriate title with each citation:

> It's useful to think of business communication as presenting a problem for which there may be no single solution (Northey, *Impact* 25).

If the author's name appears in the text, include only the title and page number(s) in parentheses. If the author's name and the title appear in the text, indicate only the page number(s) in parentheses:

> In *Impact: A Guide to Business Communication*, Northey suggests that it's useful to think of business communication as a problem for which there may be no single solution (25).

Electronic sources

In-text citations for electronic sources use the same formatting principles outlined above for print sources. Websites do not have page numbers; still, if the website you are looking at has explicit paragraph or section numbers, you may cite these instead. Note that you use a comma after the name:

> ("Stocks," sec. 2)
> (Douglas, pars. 12–15)

If your source has no page, section, or paragraph numbers, it is preferable simply to include the name of the author or title as you introduce your point.

Works cited

Your list of works cited should contain only those works you have actually referred to in the text. (Do not include works that you consulted but did not cite or directly refer to.) The following are some formatting guidelines, followed by examples of common entries in a *Works Cited* list. If the kind of source you are using isn't shown in any of the examples here, consult the *MLA Handbook*.

- Begin your *Works Cited* section on a separate page, but continue the page numbering.
- Double-space the entire list, both between and within entries and between the title and the first entry. Do not number entries, but list them alphabetically by the author's or editor's surname. If no author is given, alphabetize according to the first significant word in the title.
- Format with hanging indents: begin each bibliographic entry at the margin and indent any subsequent line five spaces.
- Separate the main divisions by periods.
- Provide page numbers for articles, chapters, and selections from PDF files. If no page numbers are given, as is true for web documents, use "n.pag."
- Indicate the medium of the resource (print, web, CD, DVD, etc.).

Book with one author

Wang, Wego. *Reverse Engineering: Technology of Reinvention*. Boca Raton: CRC, 2011. Print.

Two or more works by the same author

Entries for two or more works by the same author are arranged alphabetically by the first significant word in the title. List the author's name in the first entry only; in subsequent entries, use three hyphens followed by a period:

Chomsky, Noam. *Making the Future: Occupations, Interventions, Empire and Resistance*. San Francisco: City Light Books, 2012. Print.

---. *Occupy*. Westfield NJ: Zuccotti Park Press, 2013. Print.

Book with an editor

Relevance Ranking for Vertical Search Engines. Eds. Bo Long and Yi Chang. Waltham, MA: Morgan Kaufmann, 2014. Print.

If the book has a compiler rather than an editor, use the abbreviation "comp." rather than "ed."

Selection in an edited book

Socolaw, Robert H., and Mary R. English. "Living Ethically in a Greenhouse." *The Ethics of Global Climate Change*. Ed. Denis G. Arnold. New York: Cambridge UP, 2011. 170–191. Print.

Make sure to give inclusive page numbers for the entire piece, not just for the material you used.

Article in a journal

When listing a journal article, give the title of the article, the journal title, the volume number, the issue number if available, the year of publication, and the inclusive page numbers. Omit introductory articles ("A," "An," or "The") in the journal title:

Einstein, Albert, B. Podolsky, and N. Rosen. "Can Quantum-Mechanical Description of Reality be Considered Complete?" *Physical Review* 47 (1935): 777–80. Print.

When referencing an article retrieved from an online journal, MLA specifies that you list everything as you would for a published journal. Include the name of the database in italics, identify the medium as web, and include the date of access. If there is an issue number, include it after the volume number, separated from it by a period. Abbreviate all months except for May, June, and July:

Suleiman, Bashir M. "Estimation of *U*-value of Traditional North African Houses." *Applied Thermal Engineering* 31.11–12 (Aug. 2011): 1923–1928. *Scholars Portal*. Web. 30 May 2011.

Website

References to original content from online sources should contain the following information: the author of the content, the title of the page, the title of the site, the version or edition, the publisher or sponsor of the site, the date of publication, the medium of publication, and the date of access. Note that if no author is given, the entry should begin with the title of the page:

Statistics Canada. "Investment in new housing construction, by type of dwellings, Canada, provinces and territories, monthly (dollars)." Table 026-0017. 22 Oct. 2014. CANSIM database. 26 Oct. 2014.

MLA does not require you to list the URL unless there is no other means of locating a source.

Multimedia source

If you are referencing a multimedia source, include the description of the medium at the end of the reference but before the access date. Use n.d. if you cannot determine the original date of publication.

Shelton, Jeff. "Episode 66: Nuts and Bolts." *The Engineering Commons Podcast*. 16 Oct. 2014. Web. 21 Oct. 2014.

Chapter Checklist

☐ Keep full reference details (author, title, publisher/sponsor, dates of publication and access, medium, DOI or URL) of all the sources you consult during a project.

☐ Always confirm what style format your instructor expects you to follow so that you can satisfy the expectations easily.

☐ To simplify the preparation of your references or works cited, take advantage of the free bibliographic software your school library provides.

Note

1. IEEE. (2014) IEEE Editorial Style Manual. [Online]. Available: http://www.ieee.org/documents/style_manual.pdf

Following Conventions for Graphics and Formatting

> **Chapter Objectives**
>
> - Creating effective diagrams and other visual aids
> - Understanding when to use a table or a graph
> - Formatting equations

Introduction

Our reliance on the visual reflects the need to communicate ideas quickly with images and symbols that go beyond linguistic and cultural barriers. Illustrations capture in pictures concepts that are difficult to express in a few words. Practise your freehand drawing skills so that you can easily record ideas in sketches. Later, with the help of your computer, you can refine your sketches into detailed drawings.

When compiling material for a report, project, or presentation, consider what information you can represent graphically to illustrate important points or represent specifications. Spreadsheet programs like Excel provide options for presenting numerical data in graphs and figures. Drawing programs such as MS Paint or CorelDRAW make it easy to create, format, and annotate schematic diagrams. There are conventions, of course. Your document preparation system can make formatting automatic, but if you are producing your own material, follow these guidelines for incorporating visuals effectively:

- Information in an illustration should complement the text. Never include visuals unless they have a clear purpose.
- Simple illustrations are always better than cluttered ones. The easier it is for the reader to grasp the information quickly and accurately, the better.

- Make the caption of the visual express the point of the illustration, not just the topic. Place the caption above for tables and below for figures. Capitalize only the first letter of the first word (and of proper nouns, if any). There is room, with figures, for a sentence or two of explanation as well as a reference to the source.

- With both tables and figures, include the reference to the source directly below the graphic. For figures, place the reference directly below the caption.

- Number each table or figure so that you can make clear reference to it in your discussion. Some style manuals recommend abbreviating *Figure* as *Fig.* Note, however, that *Table* is never abbreviated:

  As Fig. 1 shows, . . .

  See Table 2 for a comparison of nozzle types.

- Introduce every visual in the text before its insertion, explaining why you have included it and what it represents, as in the example below [1]:

  Fig. 10.1 is a diagram depicting the sequence in which a message is passed among three nodes, E, T, and D. The numbers 1 to 6 indicate the order in which the message transmits: E, D, T, D, E, T, E.

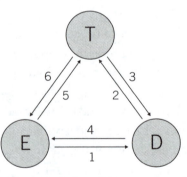

Figure 10.1   The sequence in which a message is passed among three nodes

- Prepare a list of tables and a list of figures to include after the table of contents at the beginning of a major report. Depending on how many tables and figures you have, put both lists on the same page, or give each its own page. Always include the full text of the caption in each list along with page numbers. Do not shorten or alter captions in any way.

Tables

A table can convey a large amount of information, both numerical and verbal, without losing detail. There is no simpler system for expressing comparisons.

If you are giving specific information in numerical form, a table allows you to show precise data more clearly than a graph does. This makes a table the sensible choice for data too detailed or too complex to be clearly illustrated in a graph, for example when small differences are critical, or when some or all of the information is verbal. If you have a qualification to make about one entry in a table, do so in a note by using superscript letters starting with *a* [a].

Table 10.1 illustrates the type of information that is best represented in a table. List all information in a chart format, with horizontal lines to separate headings and footnotes from the table itself, but with no vertical lines. Do not put a box around the table either.

Table 10.1 Comparison of eBook Readers

Brand	Price	Storage	Resolution	Battery Life	Weight
Nook Glowlight	$119	4 GB	212 ppi	60 days	175 g
Kobo Glo	$129	2 GB	212 ppi	30 days	185 g
Kindle Paperwhite	$189	3 GB	212 ppi	60 days	215 g
Onyx Boox T68 Lynx[a]	$199	4 GB	265 ppi	60 days	236 g

[a]shipped from UK

Note that numerous controls can decide the order of material in a table. With words, alphabetical order seems almost automatic, but in the example above, the writer has listed the brands in numerical order by cost, a meaningful distinction in a comparison. Whether you organize in alphabetical or numerical order depends on the content, but the order will always be determined by what appears in the first or second column on the left. As always, organize material from left to right and top to bottom.

Equations

Use standard equation formats, like Microsoft's Equation Editor or MathType, to produce equations, which will appear in italics. (Systems like LaTeX allow you to format equations automatically.) Unless you have only one in your text, number your equations consecutively by putting the number in parentheses flush against your right margin. You will use this number to refer to the equation in the text of your discussion:

Equation (2) reveals three parameters.

If the reference does not come at the beginning of the sentence, however, use the number alone as follows:

There are three parameters for the Black-Sholes equation as shown in (2):

$$\theta + rS\Delta + \tfrac{1}{2}\sigma^2 S^2 \Gamma = rf \qquad (2)$$

Where r is the risk-free state of interest, f is the future price of the stock, S is the current price of the stock, σ is the volatility of the stock, θ is the time rate of change of the portfolio, Δ is the weighted sum of the individual time derivatives of the stocks, and Γ is the rate of change of the portfolio's Δ with respect to the underlying asset [2].

Always be sure to define each term in the text or as a list immediately below the equation.

Figures

There are many situations in engineering where figures help a reader or audience visualize your meaning. Whether it's a document or a presentation, you can use photographs, drawings, diagrams, maps, flowcharts, or graphs to illustrate your text. Be careful to label the relevant parts of these visuals (as shown in Fig. 10.2 below)

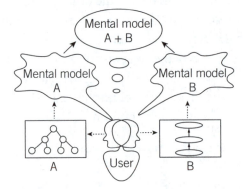

Figure 10.2   Users must reconcile information about static structure (A) and dynamic behaviour (B) into their own mental models

Source: Ian Tien and Catherine Burns [1].

and make sure that they are easy for a reader to understand. Use the caption to clarify the function of the figure briefly.

It is tempting to include photographs because of their realism, but the detail in a photograph may obscure labels or take attention away from your main point. Use photographs to provide a visual record at defined points of time (for example, photographs of a project site during its development), but use drawings or clip art instead to highlight layouts, structures, or models. Black and white is best, especially if the image is to be scanned or photocopied. If you must use colour (as in the case of electrical wiring, for example), label the colours on the diagram to avoid any confusion. In making choices about illustrating your work, always make drawings or diagrams as clean and simple as possible. In the interests of academic integrity, document the sources of all visuals in the caption unless you've created them yourself.

Graphs

Although less effective than tables in combining words and numbers, graphs blend analysis and visuals in presenting data or comparison in space and time. Software provides for a wide variety of formats and types. You are wise to limit yourself to the plainest and simplest means of making your point.

Line graphs

A line graph is best used to represent change over a period of time or space. (If it represents a distribution of data, it's called a *histogram*.) A line graph connects data points to help demonstrate or compare trends or fluctuations in trends. *Straight line graphs*, on the other hand, are visual representations of mathematical functions. If your variables are dramatically different, pick a scale that allows differences in values in data to be clear. Sometimes axes are narrow ranges; others are large. If you use a logarithmic scale, you convert your data values into their logarithms and plot these on the graph. Using logarithmic data expands the low end of the scale and compresses the high end. If you are using a log scale, remember that your measurement units cannot begin at zero. Also, be sure to label clearly whichever set of variables you have converted so that your reader is aware that the graph represents log and not linear values.

In devising a line graph:

- Put the independent variable along the horizontal axis (the *abscissa*) and the dependent variable along the vertical axis (the *ordinate*).

- Label each clearly.
- Include a legend as an inset, either with or without a border, label the axes clearly, and include tick marks to help the reader.

See Fig. 10.3 for a model of a simple line graph.

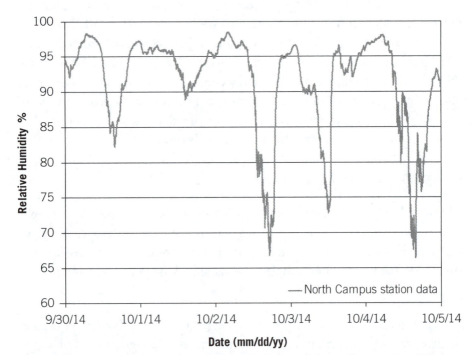

Figure 10.3 Relative humidity at North Campus from September 30 to October 5, 2014

Source: Brewster Conant Jr.

In situations where large numbers of data points are plotted on a graph, the use of lines to connect the data points would be confusing and distracting. A solution is to use a *scatter plot* (see Fig. 10.4). In scatter plots, different symbol types (diamonds, triangles, squares, circles) identify different data sources on the figure to highlight the differences between the groups. You may wish to combine similar figures into a single one to avoid repetitive captions and to facilitate comparing data on a single page, as shown in Fig. 10.4 below.

Be especially careful about clarity when you are developing a graph. Never distort it to emphasize a point—for instance, by shortening the horizontal axis and lengthening the vertical axis to make a gradual rise look more dramatic. Doing so only causes your reader to question the reliability and validity

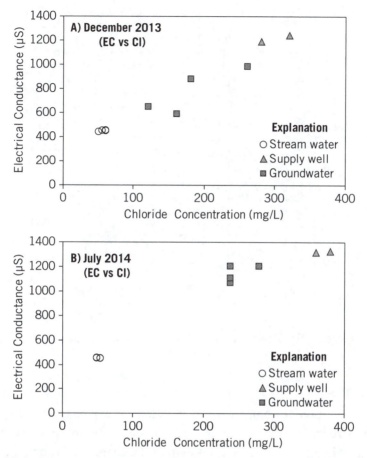

Figure 10.4   Water quality for the supply wells, stream, and groundwater
wells showing electrical conductance versus chloride in A)
December 2013 and B) July 2014

Source: Brewster Conant Jr.

of your information in interpretation—in a word, your professionalism. The
poor example below (see Fig. 10.5) features the following errors:

- It provides non-linear and uneven tick spacing along the abscissa.
- It lacks data points for Sample C.
- It lacks clear labels for both axes.
- The ordinate incorrectly displays letters vertically, not horizontally.

The graph in Figure 10.5 is unacceptable and needs to be redone.

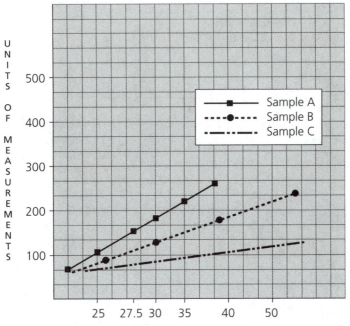

Figure 10.5 This unacceptable line graph needs revision

Bar graphs

A bar graph is used to compare elements at fixed points in time. For example, a business report might use a bar graph to show the profits made by each department in a company in a given year, or the changes in a company's sales from one year to the next. The bars can be horizontal or vertical, depending on the range of data, and they can be segmented to show different parts of the whole. For example, a bar used to show a company's sales for a series of years could be segmented to show what part of the total comprises domestic sales and what part comprises foreign sales. Bars can also be clustered or grouped to compare one category with another, as in the sample comparative bar chart in Fig. 10.6.

Pie charts

A pie chart emphasizes proportions to draw attention to the relative sizes of the parts that make up a whole. For example, it can provide a quick visual comparison of individual department sales as a proportion of total sales. It is particularly useful for highlighting expense or income categories in budgets.

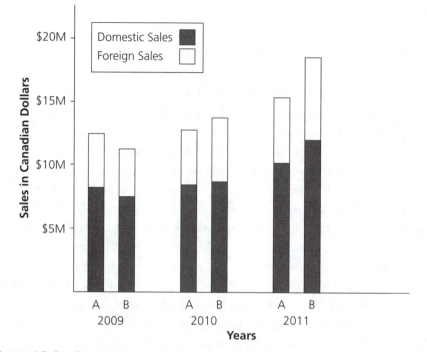

Figure 10.6 Comparison of domestic and foreign sales for companies A and B over three years

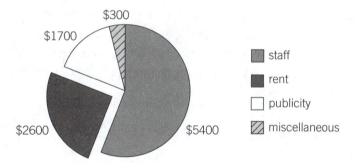

Figure 10.7 Monthly expenses for the Engineering Students Federation

With a pie chart you can also show percentages and explode one piece of the pie for emphasis, as in Fig. 10.7 above. Make sure that the shading or fill patterns you use distinguish the wedges clearly.

Tables or graphs?

When deciding whether to use a table or a graph to present data, always consider which is the most effective way of making your point. Tables have the advantage of being more exact than charts or graphs because they provide precise numerical information. On the other hand, large tables can be overwhelming and confusing in presentations if there isn't enough time to view them properly. Figures can give a more compelling impression of the overall pattern of results. That is why pie charts and graphs are more common than tables in the slides that accompany presentations.

In general, tables are a good choice if you have several sets of numbers that could get buried if you tried to list them in a paragraph. Tables help you to avoid repetitive text and long lists of numbers in a document and enhance your ability to represent and interpret the information. However, if you have data or information for a number of conditions that vary systematically, then a graph is the best way to illustrate the information and show the relationships. Given the flexibility of programs that allow multiple ways of representing information, your decision of how to illustrate your data should be based on what method conveys your information most accurately and effectively.

Remember that any illustration, even if it is created on a computer, can be manipulated. For instance, changing the range of an axis can alter the slope of a graph line to make it look steep or shallow, even when it is drawn using the same data. Likewise, axis lengths and trend lines can begin at a time that omits unfavourable periods. Although line and bar graphs are the most susceptible to distortion, the shapes and proportions of other diagrams can also give a false picture. You should always label scales on maps and drawings. Be careful to present as accurate a representation as possible so that your illustrations add to the credibility of your text.

Process Diagrams

As an engineer, you will quickly become familiar with the visual means of representing processes in a variety of formats. The controlling element of all these is time, and the main purpose of the exercise is to arrange events and activities in a sequence that acknowledges and records their interdependence. When designing process diagrams, remember to work from left to right and top to bottom. Your word processor provides the tools to make it easy to develop such diagrams.

The most familiar process diagram is the **flowchart**, which presents the steps and decisions of a standard procedure, one that can be repeated any number of times. You could use a flowchart to help you develop instructions or establish protocols for regularly repeating activities. See the chart on p. 14 for an example.

Gantt charts, such as the one appearing on p. 63 (which also establishes the context for Fig. 10.8), indicate the multiple activities involved in completing a single project. They are particularly useful in identifying the parallel or sequential activities expected of each member of a team, whether individuals, departments, or separate contractors. You develop a Gantt chart after having identified all the activities required for a project and predicting the time it will take to complete each one in context. When your project is complete, be sure to revisit your chart to note the actual timing so that you can learn from experience.

Critical Path Analysis refines the mechanics of the Gantt chart by establishing a central timeline for critical activities, those that cannot take place until another activity has been completed. In aligning these along a centre line, as Fig. 10.8 indicates, you identify the core timeline (from left to right) for the project. In identifying the interrelation of multiple tasks, the Critical Path Method allows the project manager to plan a schedule that identifies and provides flexibility for non-essential activities while keeping track of the completion of consecutive activities along the critical path. The key elements of this type of chart are the circles, which represent events, and the arrows, which represent activities. The times recorded above the arrows represent the absolute timeline for the schedule. The events that are not on the critical path may well take more or less time than allotted.

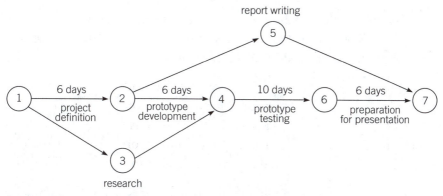

Figure 10.8   ME 100 Project: Critical Path Analysis

As you gain experience in managing projects, you will quickly appreciate the advantages of each of these methods of diagramming the amounts of time (and money and resources) a project involves. Figure 10.8 shows an example of a critical path analysis diagram.

Formatting Your Work

We say that no one should judge a book by its cover, but readers do so all the time. An attractive presentation helps make a receptive audience. Good looks won't substitute for good thinking, but they certainly enhance it. When you are presenting your work in an academic or professional context, take advantage of the following guidelines for making your reports look good:

- If you are producing a print document, use a laser printer rather than an inkjet printer and leave margins of at least 2.5 centimetres to frame the text.
- Make judicious use of formatting features such as bolding and font size for emphasis, and choose appropriate typefaces.
- Keep things simple! Be sure that the designs you are creating are not too elaborate for your purpose.

Be careful when using colour in a print document. The safest approach to ensure clarity is to use only black and white; however, colour does enhance visual appeal. Of course, your material needs to be accessible. With approximately five per cent of people colour blind, keep the following design principles in mind whenever you do use colour:

- Colour should not be the only distinguishing feature for the data points, lines, bars or shaded areas on the figure. Use different line types (solid, dashed, dotted) or different symbols (circles, squares, triangles) or different shading patterns.
- Try the photocopy test. If someone made a black-and-white copy of your figure, is it still possible to tell which line or bar or type of data is which? Some colours do not photocopy well, if at all (avoid yellow and light blue), and data would be lost if copies are ever made of your figure.
- Don't rely on reds and greens as colour choices, given the number of people who struggle to distinguish these colours.

How you format your tables and figures depends on how they will be presented to your audience. Are you preparing a written report or a visual presentation? To a certain extent, the amount of time an audience has for viewing a table or figure dictates how complicated and detailed it can be. A well-formatted table or figure suitable for a written report needs to be reformatted to improve readability (using larger fonts) for quick comprehension in a slideshow. For example, a graph with many lines on it can have a legend in a written document, but it is better to label each line on the figure in a visual presentation so that the audience is not looking back and forth to a legend to figure out which line is which.

The kinds of tables and figures discussed in this chapter are easy to create with most software. But as you have seen, while computers can help you add visual impact to your written material, they can also make it tempting to add so much detail that you obscure the basic data. Visuals are meant not to dazzle but to make your content clearer and easier to understand. Use the simplest and most accurate means available to make your point.

Finalizing your work

Word processors have many refinements for making the best product possible, but do not trust them too far. For example, your word processing software catches typing as well as spelling mistakes, but it can't point out actual words that are wrongly used (such as "there" when you mean "their," or "trail" when you thought you typed "trial"). A spell checker is a useful first check, never a final one. Similarly, grammar checkers may pick up on some grammar and stylistic problems, but they are often wrong. They cannot match the judgment of a good human editor.

For final editing, most experienced writers suggest working from the printed page rather than from the screen. The same is true for slideshows. Print one draft to read and edit before producing the final version. Your eye gets a better sense of the text by seeing one page at a time than by scrolling on a screen. Read over your work with a critical eye, knowing that you can easily change something that is unsatisfactory.

When you are preparing the final document, resist the temptation to use colour, fancy fonts, or unnecessary clip art because too much is distracting not enhancing. Keep in mind that while a computer can make your work look good, fancy graphics and a slick presentation never replace intelligent thinking. Remember that your computer is a tool—no more.

Protect Your Work

Lost files are a nightmare. If you keep a copy of every major draft of your work, you need never worry about losing your work forever. Save each successive file with its name and a version number (e.g., filev1, filev2) so that you can easily recover work if you decide you don't like your most recent changes. Doing so also provides a traceable chronology in case your authorship is questioned. Always have an additional copy besides the one on your hard drive. If you send yourself an electronic copy, you'll have access to it wherever you go. Most importantly for you as a student, retain a copy of every file at least until you receive your final grade for the course. Getting into the habit of following these steps is a good investment in both time and security.

Chapter Checklist

☐ Evaluate your data carefully to determine the most effective means of presenting information visually.
☐ Introduce tables and figures by name in your document before inserting them in your text.
☐ Keep graphs, charts, and diagrams simple and label them clearly.
☐ Follow conventional formatting and design principles.

Note

1. I. Tien and C. Burns, "Flowlinks: A technique for merging static and dynamic mental models of software systems," unpublished paper.

Writing Tests and Examinations

Introduction

Most students feel nervous before tests and exams. It's not surprising. Writing an exam, even if you are allowed to bring a cheat sheet or a textbook, imposes special pressures because both the time and the questions are restricted. With some exams worth 50 per cent or more of the grade for a course, you want to be prepared to answer every question thoroughly and accurately. On the surface, objective tests may look easy because you don't have to develop extended answers, but they force you to be decisive about your answers. To do your best, you need to feel calm—but how? The following general guidelines should help you approach any test or exam with added confidence.

Preparing for the Exam

Review regularly

Exam preparation begins long before the exam period itself. A daily or weekly review of lecture notes and texts helps you remember important material and relate new information to old. If you don't review regularly, at the end of the term you'll be faced with relearning rather than remembering.

Set memory triggers

As you review, condense and focus your material by writing down in the margin key words or phrases that generate whole sets of details in your mind. The trigger might be a concept word that names or points to an important theory or definition, or it might be a quantitative phrase such as "four elements of a feasibility analysis" or "six types of intellectual property."

Sometimes you can create an acronym or other memory device that triggers an otherwise hard-to-remember set of facts. (Remember how useful BEDMAS was when you were first learning to do calculations?) Since the difficulty of memorizing increases with the number of items you are trying to remember, any method that reduces this number increases your effectiveness.

Ask questions: Try the three-C approach

Think of questions that will get to the heart of the material and force you to examine the relations between various subjects or issues; then think about how you would answer them. The three-C approach, discussed on pp. 81–83, may help. For example, reviewing the **components** of the subject could mean focusing on the main parts of an issue or on the definitions of major terms or theories. When reviewing processes, you might ask yourself what the causes or results of those **changes** are. To review **context** you might consider how certain aspects of the subject—issues, theories, actions, results—compare with others in the course. Essentially, the three-C approach forces you to look at material from different perspectives.

Copies of previous exams are useful both for seeing the types of questions you are likely to be asked and for checking the thoroughness of your preparation. Your professor will tell you where to find old exams if they are available—check your departmental library or course websites for ideas. It's also a good idea to work on problem sets with friends taking the same course. Just remember that the most useful review questions are not the ones that require you to recall facts and formulas but the ones that force you to analyze, synthesize, and apply what you have learned.

Allow extra time

Give yourself lots of time to get to an exam. Nothing is more nerve-wracking than the fear of being late. If you have to travel by car or public transit, don't forget that traffic can jam, as can alarm clocks. Always allow for a margin of error. The same advice applies if you are writing a test electronically on your course's learning management system. Don't wait until the last minute to submit your answers, or you may find the drop box closed.

Of course, if the inconceivable does take place and you are late for or miss an exam, document the situation as best you can (for example, with a police report or emergency-room record). Let both your professor and the department know about the situation immediately so that alternative arrangements may be made.

Writing an Exam

Read the exam

Much as time seems an issue, an exam is not a hundred-metre dash. Instead of starting to write immediately, take time at the beginning to read through the questions and create a plan. A few minutes spent thinking and organizing at the outset brings better results than the same time spent scribbling furiously. It's essential to read each question carefully to make sure you understand exactly what's being asked. Nothing is more frustrating than comparing notes with a classmate afterwards and discovering that you misunderstood instructions.

Apportion your time

Read the instructions carefully to find out how many questions you must answer and to see what choices you have. Subtract five minutes or so for planning, and then divide the time you have left by the number of questions you have. If possible, allow for time at the end to review your answers. If the instructions on the exam indicate that not all questions are of equal value, apportion your time accordingly.

Choose your questions

Decide which questions you will do and in what order. You never have to do an exam in the same order as the questions. Tackle first the questions you can answer fastest and best, to save time for those you need to work hard at answering well. That way you will be sure to get the maximum marks.

Keep calm

If your first reaction on reading an exam is to panic, force yourself to be calm, take slow, deep breaths to relax, and then decide which question you can answer best. Even if the exam looks impossible at first, there's always one question that seems manageable. That's the one to begin with, to get you going and increase your confidence. By the time you have finished the first answer, you will probably find that your mind has automatically begun to consider the next question.

Read each question carefully

As you consider each question, read it carefully and identify key words. The wording specifies the direction your answer should take. Be sure that you don't overlook or misinterpret anything (something that's easy to do when you're nervous). If you are required to complete the exam on the test paper itself, note the amount of space allotted for each answer. It tells you better than anything else how much detail your answer should contain.

Note the difference between questions and instructions in the phrasing. Questions beginning *Do, Does, Is*, or *Are* call for a straightforward answer of *yes* or *no* or some similar choice. The *Why, How*, and *What* questions call for justification, explanation, and details—all expressed as short answers, preferably in sentences. It's easy to recognize exam questions in this category by their end punctuation (?).

Other exam questions are worded as instructions to be followed precisely. You will recognize, for example, the standard verbs that direct you to do calculations in order to solve problems: *calculate, compute, determine, estimate, find*, and *solve*. Others require some kind of illustration or graph: *chart, draft, draw, diagram, highlight, plot, sketch*. Short answers that depend on recall are prompted by the following: *enumerate, give, list*.

In questions that call for extended discussion, the verb again determines the approach to take in answering. For example, note the instructions implicit in the following:

analyze   divide into components; explain features; describe structures, methods, or operations.

classify   divide into groups according to shared characteristics. (See pp. 7–8 for more detail.)

compare   identify differences as well as similarities. (See p. 8 and p. 53 for more detail.)

define   tell what something is, or how it works. (See p. 52 for more detail.)

describe   present features, characteristics, details, perhaps including a sketch, graph, or diagram. (See p. 7 for more detail.)

discuss   examine or analyze in an orderly way. This instruction allows for considerable freedom of approach, as long as you take into account contrary evidence or ideas.

> *evaluate* analyze strengths and weaknesses, providing an overall assessment of worth.
>
> *explain* show how or why something happens.
>
> *identify* describe, illustrate, explain, or present examples.
>
> *outline* identify steps or stages, without development.
>
> *show* demonstrate or prove, mathematically, scientifically, or otherwise logically.
>
> *trace* review chronologically, listing causes, effects, or stages or steps in a process.

These and other verbs tell you how to handle your answer. Pay attention to them right from the beginning of the test.

Make notes

As a preliminary strategy, jot down key ideas, details, and formulas on rough paper or the unlined pages of your answer book. These notes will keep you from forgetting something as you write. Use them to organize yourself once you start to answer in earnest.

Be direct

Get to your points quickly, adding examples and details as necessary. It's best to use a direct approach, backing up each general statement with specifics. Don't add unnecessary qualifications or elaborations (especially on objective exams), as they suggest a lack of confidence or knowledge. Remember that your exam is one of many to be marked by someone who has to work quickly, so the clearer your answers are, the better they will be received. For each answer, give details that prove you really know your material. General statements show you can assimilate information, but you get top marks for specifics—as long as they are accurate as well.

Write legibly

Poor handwriting makes markers cranky. When the person correcting your exam has to struggle to read your answers, you may get lower marks than you deserve. If for some special reason (such as a physical handicap) your writing is hard to read, you may be able to make special arrangements to use a computer. If your writing is just not very legible, it's probably better to print. When you are taking an essay exam, it's a good idea to write on every second or even every third line of the booklet. Doing so not only makes your answers easier to read but also leaves you space to make revisions and additions if there is time later on.

Keep to your time plan

Keep to your plan, and don't skip any questions. It's usually possible to score part marks for a question you can't fully answer, so put something down for every question if you can. If you find yourself short of time, summarize what you have written and move on. If you have time, you can always go back to complete an answer once you've answered the more valuable questions to the best of your ability. If you ever change an answer, cross out the old one neatly. This way, you won't have a lot of rewriting to do if you decide later that you prefer your first answer.

Reread your answers

No matter how tired or fed up you are, try to leave time to review your answers at the end of the exam. Check especially for readability. Revisions that make answers easier to read are always worth the effort.

Writing an Open-Book Exam

Don't think that permission to take a cheat sheet or textbook into the exam room guarantees success. Do not fall into the trap of relying too heavily on your reference material. It's easy to spend so much time rifling through pages and looking things up that there isn't time to produce thorough or accurate answers. The result may be worse than if you had been allowed no aids at all.

If you are allowed to bring in a cheat sheet, be careful to follow directions exactly. Normally, you will be given a definition of what you're allowed to bring—for example, a standard-sized sheet of paper with writing on one side only. It makes sense to follow such instructions to the letter; otherwise, you might find yourself "sheetless" when it matters most.

In the exam itself, use your aids only to check numbers and look up specific, hard-to-remember details. For instance, if your subject is biotechnology, you can look up genetic formulas. For a statistics test, you can apply t-tests and chi-squared tests to given data. For an exam in thermodynamics, you can refer to criteria for equilibria and definitions for entropy. You can check classic references and find exact definitions of key concepts—as long as you know where to find them quickly. In other words, use the aids to make your answers precise and well illustrated. Don't expect to use them to reduce the time you spend studying and preparing.

Writing a Take-Home Exam

If you are given a take-home exam, you have time to plan your answers, consult your texts and other sources, and even work out answers with your classmates. The catch is that you probably still have less time than you would for an ordinary assignment. Use your time to produce a professional response to the exam question(s). After all, a 24- or 48-hour time limit is no different from a typical business situation, with a client demanding instant results. Keep in mind that you were given the exam to test your overall command of the course material. Your reader is seeking evidence that you have understood and absorbed the principles presented in class.

One reason an instructor may have for assigning a take-home test is to present you with a complicated problem that is impossible to answer in two or three hours. In this case you're being tested on your efficiency in assessing a problem and developing a solution over time. Take as much time as necessary to analyze the demands of the test question so that you can offer a considered response, one that stands up to scrutiny.

The guidelines for a take-home exam are similar to those for a regular exam. The only difference is that you don't need to keep such a close eye on the clock:

- Unless told not to, use your word processor to provide the professional appearance that the time allows.
- Organize your answer so that the reader can easily follow it. Use headings if they are helpful.
- Use specifics (detailed diagrams, algorithms, equations, or graphs) to back up your points.
- Where possible, show the range of your knowledge of course content by referring to a variety of examples rather than the same old ones.
- Focus on showing that you can analyze and evaluate material—in other words, that you can do more than simply repeat information.
- If you are to upload your exam through the course management system drop box, make sure you begin the process at least ten minutes before the deadline to allow for glitches.

Writing a Multiple-Choice Test

Multiple-choice tests are particularly common in courses with large numbers of students, for they make marking manageable. The main difficulty with these tests is that the questions are often designed to confuse students who are not certain of the correct answers. If you tend to second-guess yourself or if you are the sort of person who sees alternatives to every question, you may find such tests particularly difficult at first. Fortunately, practice almost always improves performance.

Preparation for multiple-choice tests is the same as for other kinds of exams that emphasize your ability to recall or apply information. Although there is no sure recipe for doing well, other than a thorough knowledge of the course material, the following suggestions may help.

Look at the marking system

If marks are based solely on the number of right answers, pick an answer for every question, even if you aren't sure it's the right one. For a question with four possible answers, you have a 25 per cent chance of being right—even if you pick the answers at random.

On the other hand, if there is a penalty for wrong answers (with marks deducted for errors), make sure you guess only when you are fairly sure you are right, or when you are able to rule out most of the possibilities. Making wild guesses in these circumstances does more harm than good.

Do the easy questions first

Go through the test at least twice. On the first round, deal only with those questions that you can answer without difficulty, and don't waste time on troublesome questions. If all the questions are of equal value, start by getting all the marks you can on the ones you find easy. Save the more difficult questions to tackle the second time around. This approach has two advantages: first, you won't be forced by time to leave out questions that you could easily have answered correctly; second, when you come back to a harder question on the second round, you may find that you have figured out the answer in the meantime. Sometimes, reading other questions and answers reminds you of material you had forgotten.

Make your guesses educated ones

If you have to guess, at least increase your chances of getting the answer right. Forget about intuition, hunches, and lucky numbers. More importantly, forget about so-called patterns of correct answers—the idea that if there have been two "A" answers in a row, the next one can't possibly be "A" as well. Many test-setters either don't worry about patterns or else deliberately elude pattern-hunters by giving the right answer the same letter or number several times in a row.

Remember that constructing good multiple-choice tests is a special skill that not all instructors have mastered. In many cases, the questions they pose, though reasonable enough as questions, do not produce more than one or two realistic alternatives for answers. In such cases, the test-setter may resort to some less realistic options—which you can spot if you keep your eyes open. James F. Shepherd [1] has suggested a number of tips that will increase your chances of making the right guess:

- Start by weeding out all the answers you know are wrong, rather than looking for the right one.
- Avoid any terms you don't recognize. Some students are taken in by anything that looks like sophisticated terminology and may assume that such answers must be correct. In fact, these answers are usually wrong (the unfamiliar term may well be a red herring, especially if it is close in sound to the correct one).
- Avoid extremes. Most often the right answer lies in between. For example, suppose that the answer choices are the numbers 800,000; 350,000; 275,000; and 15: the highest and lowest numbers are likely to be wrong.
- Avoid absolutes, especially on questions dealing with people. Few aspects of human life are as certain as is implied by such words as *everyone*, *all*, or *no one*; *always*, *invariably*, or *never*. Statements containing these words are usually false.
- Ignore jokes or humorous statements.
- Choose the best available answer, even if it is not indisputably true. Choose the long answer over the short (it's more likely to contain the detail needed to make it correct) and the particular statement over the general (generalizations are usually too sweeping to be true).
- Choose "all of the above" over individual answers unless you can immediately spot a wrong answer. Test-setters know that students with a patchy knowledge of the course material often fasten on the one fact they know. Only those with a thorough knowledge recognize that all the answers listed are correct.

One Final Tip

If you have time at the end of the exam, go back and reread the questions. One or two wrong answers caused by misreading can make a significant difference to your score. However, don't start second-guessing yourself and changing a lot of answers at the last minute. Studies have shown that when students make eleventh-hour changes, they are often wrong. Stick with your original decisions unless you know for certain that you have made a mistake.

Chapter Checklist

- ☐ To prepare for exams effectively, review course material regularly, paying attention to the kinds of questions your instructor raises and addresses in class or to those appearing on previous exams.
- ☐ Take the time to read exam questions slowly and carefully, making sure you understand what you're expected to do.
- ☐ If you finish an exam before time is called, take the time to go over your answers rather than leave early.
- ☐ Go over graded exams carefully in preparation for the next one.

Note

1. J.F. Shepherd, *College Study Skills*, 6th ed., Boston: Houghton Mifflin, 2002.

12

Writing a Resumé and Letter of Application

Chapter Objectives

- Preparing different types of resumés
- Writing cover letters

Introduction

Whether you are looking for a co-op placement, an apprenticeship program, a summer job, admission to graduate school, or permanent employment, you will have to produce a resumé and a cover letter. Both are intended to open the door to the next stage in the job hunt: the interview. You may even need these to apply for admission to some courses, especially those with work-study options. The person who reads your application has no time to waste, so you need to be brief yet precise, personal yet professional. Check with the career services advisors at your college or university for models that are appropriate for students like you.

Preparing a Standard Resumé

Think of a resumé as more than a summary of facts. Consider it a marketing tool, designed to show how your talents and experience match an employer's needs. You can start with a standard template, but how you organize it and which details you emphasize always depend on the position you're applying for. First, take the time to do research to educate yourself about the activities undertaken by your potential employer. Everything is likely available online: begin with the employer's homepage, and link to sections describing history, organizational structure, services, products, and future plans. Then take the description of the position you seek, and match it to your background and skills. Once you've completed this preliminary work, you can begin to develop a resumé that's tailor-made.

The resumé itself begins with your contact information, after which you should summarize in four to six bullet points those qualifications that make you a particularly suitable candidate; putting them first ensures that these points will be noticed immediately. Then present, from most to least recent, your education, your experience, and your other skills.

Whatever format you choose, your goal is to keep the resumé concise (no more than two pages) while including all the information that helps you sell your skills. A reader quickly rejects a resumé that seems sloppy or looks padded. Still you want to list all the activities, experience, or skills that can represent important attributes, such as adaptability, a sense of responsibility, or even a willingness to work weekends. For example, studying abroad or volunteering for a student organization is worth mentioning, especially if you can highlight skills you learned as a result.

The sample resumés that follow offer good examples of appropriate content and formats. Within the education, awards, and experience sections, use reverse chronological order to feature the most recent item at the beginning.

- **Name.** Begin with your name in a font that's one size larger than the one you use in the rest of your resumé. If you go on to a second page, insert your name at the top as a running head followed by the page number.
- **Contact information.** This includes your home address, phone numbers, e-mail address, and website, if you have a professional one that presents samples of your work. If you have a temporary student address, remember to indicate where you can be reached after the term is over. For your own credibility, use your school e-mail address rather than a social one. It is also wise to make sure that your voice-mail message is short and professional, to ensure that a prospective employer will even leave you a message.
- **Job objective** (optional). It may be helpful to remind a hiring committee of your current aim for employment (if it's a short-term position, it will normally be different from a career objective). Without naming the employer, use the job name provided in the ad.
- **Summary of qualifications.** This section lists the most significant experience and skills that qualify you for the position. The first bullet should summarize the number of years (or work terms) of work experience most directly connected to the job, and the remaining bullets will present your skills, abilities, and special knowledge as they apply

to the job specifications. Include computer knowledge, leadership skills, and personal characteristics related to work (for example, ability to meet tight deadlines, problem-solving skills, and fluency in another language). If they are current and relevant, list such credentials as certificates in lifesaving or the handling of hazardous materials.

- **Education.** Include all of your degrees and related diplomas or certificates, along with the institutions that granted them and the years. If it will help your case, you can also list courses you have completed that are relevant to the job.

- **Academic awards or honours.** If you have three or more, list them in a separate section along with the years they were awarded. Include them with your education details if you have only one or two, especially if they are no longer recent.

- **Work experience.** Give the name and location of your most recent employers, along with your job title and the dates of employment. Instead of outlining duties, which any employee may or may not carry out well, list your accomplishments, using point form and action verbs. Pay special attention to parallel phrasing, making sure every bullet works with "I" as the subject (see p. 200). Here are two examples:

  - Designed, administered, and reported on a resource management survey
  - Supervised a three-member field crew

If you have worked as a research or teaching assistant, be sure to state the type of work you did and the name of your supervisor. Include such details, either in the chronological listing of previous employment or in a separate category, this way:

  - May–August 2014. Research assistant for Professor Marika Szabo, independently completing 10 experiments and reporting the results in a paper entitled "The Liquid Limit of Leda Clay," Carleton University, Ottawa.

If relevant, include volunteer experience, clearly indicated as such, either as a separate section (if you have a lot in addition to paid positions) or as part of your list of work experience.

- **Extracurricular activities (optional).** If there is still room on your second page, take this opportunity to list extracurricular expertise not mentioned in your Summary of Qualifications. Naming a few achievements or interests, such as athletics, music, or travel, confirms that you are well rounded or especially disciplined. Avoid listing general activities that show passive or minimal involvement. For example, while reading is a commendable activity, including it says nothing about your versatility as an employee. Include only those interests that set you apart from the average applicant.

- **References** (optional). It is no longer necessary to write "References available upon request" at the end of a resumé. Either you will have been asked to include them (which you do on a separate sheet), or you will be expected to bring them with you to an interview. As a courtesy, always confirm a person's willingness to give you a strong recommendation before using his or her name. Then give the full name, complete title, address, phone number, and e-mail address of each one.

Standard resumé

Raj Choudhury

Local Address (until 1 May 2015):

123 College Rd

Halifax NS, B3H 2R2

Phone: 902/323-4567

Fax: 613/456-0001

Email: rchdhry@dal.ca

Permanent Address:

45 Main St

Ottawa ON, K1K 1K4

Phone: 613/456-0000

www.rajchoudhury.ca

Job Objective: An entry-level position as junior software developer

Summary of Qualifications:

- Six years of experience in electrical and software engineering at school and on the job.
- Extensive computer skills. Operating Systems: Windows XP and Windows 7, Solaris, Linux. Languages: C/C++, SQL, HTML, DHTML/CSS, and JavaScript. Applications: XEmacs, Matlab, AutoCad, SolidWorks, Cygwin, Eagle and Eagle PCB layout.
- Strong background in tech support. Took courses in integrated circuits and high-speed PCB design and used these skills on the job with Texas Instruments.

- Proven ability to follow instructions and procedures with minimal supervision.
- Excellent written and oral communications skills.

Education:

- MEng in Electrical and Computer Engineering, Dalhousie University, expected in June 2016 (Thesis: "Machine Architecture and Linear Memory")
- BEng in Electrical and Computer Engineering, University of Victoria, 2013

Honours and Awards:

- Dean's List, 2009–2013, University of Victoria
- NSERC Industrial Undergraduate Research Award, 2012

Work Experience:

May–August 2013 – Software Engineer for Texas Instruments, Toronto

- Performed software validation and provided quality assurance
- Designed a multi-headed debugging driver with Visual Studio C++

May–August 2012 – Systems Engineer for Ford Motor Company in Windsor, Ontario

- Wrote, modified, integrated and tested software code for e-commerce and other internet applications.
- Provided tech support and trouble shooting for three departments.
- Trained a college co-op student to take over plant website maintenance.

May–August 2010, 2011 – Research Assistant for Path Technologies, Gibson, BC

- Performed daily and weekly lab tests and prepared reports of results.
- Maintained and updated data base.

Extracurricular Activities:

- Certified in Standard First Aid & 2-Person CPR, WHMIS, and PADI Open Water Scuba diving
- University of Victoria Engineering Society Director, 2011–2012
- Member of University of Victoria varsity basketball team, 2009–2011

Applying For a Position

Most employers post job opportunities either on their organization's website or on standard sites like Workopolis or Monster, and the Government of Canada maintains a master list of positions on Job Bank. E-mail applications are encouraged because of their timeliness. However, there are pitfalls for the unwary. These guidelines will help you avoid them:

- To ensure that your message is not designated as spam, use the subject line to identify the job exactly as it is posted, for example, "Process Engineer - JO189."
- Whenever possible address your message to a specific person whom you do not address by first name ("Dear Ms. Lee" rather than "Hi Karen!"). If a name has not been linked to the e-mail address listed with the job description, check the company's electronic directory. Use "Attention: Human Resources" as a last resort.
- Keep the text of your e-mail brief. It has three aims:
  - Name the position and indicate your interest in it.
  - Indicate that your cover letter and resumé are attached.
  - Thank the reader for considering your application.
- Scan your letter and resumé as separate attachments rather than put them in the body of the e-mail message. That way you ensure the format will appear to the reader exactly as you intended. The appearance of e-mail can go visually askew at the receiver's end, corrupting a neatly formatted message. An attachment, by contrast, will look the way you created it and can easily be printed. To confirm that you recognize the conventions of applying for a job, always attach a signed cover letter, even if you have included some of the text as the body of your e-mail.
- Before pressing the "send" button, proofread your e-mail message carefully. A job application is too important for you to run the risk of looking sloppy. Pay special attention to the spelling of people's names.

Preparing a Scannable Resumé

Many employers add resumés to databases in case an applicant is qualified for other positions with their organizations. To prepare your resumé so that it can be scanned for inclusion in a database or for electronic posting, take note of the following modifications to the standard printed format:

- **Presentation.** Minimize formatting features so that your resumé is easy to scan. Use only two font sizes (the larger for headings and the smaller for contents), and line everything up against the left margin. Use punctuation only when necessary to keep items separate, and keep lines as short as you can (no more than 70 characters, including spaces).
- **Keywords.** After listing your name and contact information, include a short list of words for which your resumé can be scanned. Supply nouns that refer to your qualifications and expertise as well as to the job you are seeking. (Refer to similar job descriptions or ask your school's career advisor to help you make the best choices.)

Scannable resumé

The following example gives the first few lines of the standard resumé that appears on pp. 150–151 revised to suit an electronic (scannable) format:

choudhuryraj.txt

RAJ CHOUDHURY
45 Main St
Ottawa ON K1K 1K4
Ph 613/456-0000
Fax 613/456-0001
choudhury@intranet.ca

KEYWORDS:
electrical and computer engineering, software engineering, debugging, website maintenance, tech support, systems, research

POSITION DESIRED:
Software developer

SUMMARY OF QUALIFICATIONS:
Six years of experience in electrical and software engineering at school and on the job

Extensive computer skills. Operating Systems: Windows XP and Windows 7, Solaris, Linux. Languages: C/C++, SQL, HTML, DHTML/CSS, and JavaScript. Applications: XEmacs, MATLAB, AutoCad, SolidWorks, Cygwin, Eagle and Eagle PCB layout.

Preparing a Functional Resumé

A resumé with traditional divisions is not the only kind that works. If you want to emphasize your abilities and your versatility in addition to the jobs you've had, prepare a **functional resumé** instead. A functional resumé has categories for experience in different areas of expertise (for example, research, field work, administration, sales). Or it may focus on personal attributes such as initiative, teamwork, analytic ability, or communication skills. As always, tailor your selection to match the requirements of the position you are applying for.

Functional resumé

CAROLINE PAGE

16 Harbour St
Toronto ON M4M 1V6
416/491-3020
cnpage@sympatico.ca
www.cnpage.ca

Objective: Work as an environmental consultant in wastewater management.

Profile: A highly organized environmental engineer (B.A.Sc., University of Waterloo, 2015) with a positive attitude and consistent ability to meet deadlines in challenging, high-pressure surroundings.

Environmental Engineering:

- Developed MATLAB code for modelling atrazine transport in an aquifer (2014).
- Assisted in site characterization and site selection for active and passive water treatment systems (2014).
- Conducted an audit of a paper mill freshwater system (2013).

Field Work:

- Conducted water and sediment sampling at a mine site (2013).
- Planted a constructed wetland for treatment of wood-waste landfill leachate (2013).

Laboratory and Analysis:

- Identified, prepared, and analyzed specialized bacterial cultures for wastewater treatment (2014).

- Conducted thermogravimetric analysis (TGA) and direct-current plasma (DCP) analysis of samples (2014).
- Carried out experiments using a variety of biotechnology lab techniques (aseptic technique, culture growth, enumeration, gram staining, etc.) (2013, 2014).

Communication Skills:

- Fluent in French, both written and spoken
- Part-time teaching and research assistant at the University of Waterloo (January 2013 to present)
- Instructor/Counsellor for Engineering Science Quest (a summer camp for pre-teen amateur scientists) (Summers 2011, 2012)

Software Tools:

- Skilled in Aspen-HYSYS, including extensibility
- Proficient in MSOffice and Visual C++; working knowledge of MATLAB and VB

Other Achievements & Activities:

- Six-time recipient of the UW Engineering Upper Year Scholarship for academic excellence
- Member of the Dean's Honours List from 2010–2015
- Musician: guitar and Celtic whistle
- Avid cyclist and runner
- Interested in environmental remediation and international development

Choosing the Best Content

The content of a resumé should be professional, so never draw attention to weaknesses you may have, such as lack of experience. Be sure to tailor your list of special skills to suggest a fit with each job you apply for, so that the reader can see at a glance that you meet the job requirements. Finally, never claim more for yourself than is true. Including false information in a resumé is grounds for dismissal if it is discovered.

Keep in mind that you are not required to state anything about your age, place of birth, race, religion, or gender. To be sure that your application will be considered, of course, you must provide all the information requested when you are completing an application form that has set questions.

Writing a Cover Letter

Most employers judge the writing of the letter that accompanies a resumé. To catch their attention, make sure your letter is specially drafted with the company in mind. Don't use the exact same letter for all applications. Instead, each cover letter will follow a standard pattern. In terms of content, what matters is *not* that you want a job but that the employer is looking for someone just like you. Use the **Summary of Qualifications** to help you craft your letter, but adjust the wording so that you don't simply repeat what's in your resumé. Make each paragraph fulfill a specific function:

1. Identify the position and state your interest in applying for it.
2. State your best qualification and refer the reader to your resumé.
3. State your next best qualifications.
4. Politely state your availability for an interview.

The key is to link your skills to the position, not simply state information.

One challenge in writing a letter of application is to tell your reader about yourself and your qualifications without seeming too self-focused. Two tips can help:

1. Limit the number of sentences beginning with *I*. Instead, try burying *I* in the middle of some sentences, where it will be less noticeable, for example, "For two months last summer, I worked as a . . ."
2. Avoid, as much as possible, making unsupported, subjective claims. Instead of saying "I am a highly skilled manager," offer specifics like, "Last summer I managed a $90,000 field study with a crew of seven assistants." Rather than claim "I have excellent research skills," you might say, "Based on my volunteer work with first-year students, Professor Anson Park hired me as a lab assistant for CIVE 221 (Advanced Calculus) in Fall 2014."

Here is an example (to be imitated, *not* copied) of an application letter that relates the applicant's background to the current needs of a company:

1 April 2015

Steven Nazar, Personnel Director
Outlands Developments
110 Duplessis Blvd
Ottawa, ON N5N 1T7

Dear Mr. Nazar:

Your advertisement in the *Ottawa Citizen* for a Junior Environmental Officer caught my attention, since my qualifications match those you are seeking.

As a student graduating with a BEng in Environmental Engineering from the University of Regina, I would like to apply for the position.

Beyond my specialist academic program, I have had relevant environmental field experience. As my resumé indicates, I have spent two summers working with Professor Susan LeClerc in her studies of wetland pollution south of James Bay, helping both in the laboratory and in the wetlands.

I also enjoy and am used to the kind of outdoor work you require. In my first summer as a university student, I worked as a tree planter in harsh conditions in northern British Columbia. If there is work available in eastern Canada, I am able to speak French well enough to converse with your French-speaking clients in Ontario and Quebec.

I would appreciate the chance to discuss with you how I could contribute to Outlands Developments. Please call or e-mail me to arrange an interview at your convenience. I look forward to hearing from you.

Sincerely,

Marco Garon

Marco Garon

Making a Good First Impression

When you apply for a job, your application will be one of many. Of course, it must pass a screening process before you will be invited to an interview. For this reason, it is essential that you submit a package that looks as professional as you will when you present yourself in person. Your application package will be judged not just by what you say but also by how you say it. That's why you must take the time to double-check for grammar and spelling errors and to verify that your application documents are well formatted.

Be certain that you have spelled all names correctly. To give yourself the best chance to be asked in for an interview, take as much time as you need to make everything perfect.

With job interviews, as with applications, first impressions count. When you are asked to an interview, you want to confirm that you are as professional as your resumé promised. Take advantage of your school's career centre advisors to help you prepare, and remember that body language and tone of voice are as important as the appropriateness of the answers and details you provide at your interview. Then, if everything comes together just right, you can look forward to a job offer that's well-deserved.

Chapter Checklist

- ☐ Take advantage of career services advisors and writing centres to help get you started.
- ☐ Research both the position you want and the company it's with.
- ☐ Match your skills and experience to the job description and prepare your Summary of Qualifications accordingly.
- ☐ Summarize your education, awards, and experience in reverse chronological order followed by skills and extracurricular activities.
- ☐ Prepare the rest of your resumé.
- ☐ Write a cover letter that introduces you and your best qualifications.
- ☐ Write an e-mail message to the hiring committee and attach your resumé and letter.
- ☐ Proofread slowly and carefully all documents at least twice before sending.

Writing for Readability

13

Chapter Objectives

- Writing in a clear, concise, and convincing manner
- Staying consistent and objective in your writing
- Making important ideas stand out

Introduction

Technical writing demands both objectivity and professionalism. Your professional success depends on your ability to persuade your reader to accept your ideas. Whether you are reporting a straightforward procedure or making recommendations, your writing must always be easy to read and to understand. There is nothing unprofessional about writing that's simple and to the point.

Sentence length and style influence a reader's response to writing. The language you use to convey your ideas must never take attention away from those ideas. Studies of user-friendly language indicate the obvious: short words and short sentences are generally easier on readers, even sophisticated readers. People who do a lot of reading like it to be easy.

Because of its technical content, scientific writing is characterized by a specialized vocabulary and academic tone that make it sound complicated. Novice writers often feel that they must adopt a similarly elaborate style. Not true. To convince your reader that you are in control of your subject, make your sentences precise and economical, not dense or convoluted. The most effective and readable style is one that is clear and concise, confident and consistent. When editing, you should always look for ways to make it easy for the reader to understand exactly what you mean.

The guidelines in this chapter apply to the final stages of the writing process, when you already have a rough draft in hand. As you examine that draft, consider ways of reducing the overall length as well as varying the average sentence length. In general, the shorter the sentences are, then the easier the text is to read.

Be Clear

Use plain English

Plain words rather than flowery ones almost always make your sentences more readable. Many of our most common words—the ones that sound most natural and direct—are short. By contrast, most words derived from other languages (like French or Latin or Greek) are longer and more complicated. Given the number of synonyms and choices, you should be wary of words loaded down with prefixes (*pre-*, *post-*, *anti-*, *pro-*, *sub-*, *maxi-*, etc.) and suffixes (*-ate*, *-ize*, *-tion*, etc.). Too much dependence on big words makes your writing hard to read. If you can substitute a short word for a longer one, do so:

Flowery	*Plain*
accomplish	do
amongst	among
cognizant	aware
commence	begin, start
conclusion	end
concur	agree
determinant	cause
efficacious	effective
endeavour	try
fabricate	build, create
finalize	finish, complete
firstly	first
initiate	begin
maximization	increase
modification	change
numerous	many
oration	speech
proceed	go
rectify	correct
remuneration	pay
requisite	needed, necessary

subsequently	later
systematize	order, arrange
terminate	end
transpire	happen, take place
utilize, utilization	use
whilst	while

Suggesting that you write in plain English does not mean that you should never pick an unfamiliar, long, or foreign word, especially if it is the only one that expresses a complicated concept concisely (for example, there is no simpler way to refer to *bioremediation* or *complementarity*). But do not needlessly clutter your sentence with longer words or phrases when shorter alternatives exist. Note how big words and long phrases can cloud rather than clarify meaning, as in the following example:

orig. The addition of the acid and the subsequent agitation of the solution resulted in the formation of crystals. (18 words)

rev. Crystals formed when the acid was added and the solution shaken. (11 words)

If you find yourself selecting words or phrases only because they look impressive, you may find your writing criticized for sounding awkward instead.

Choose clear wording

Count on the dictionary to help you understand unfamiliar or archaic words or technical senses of common words. A good dictionary will also help you develop your vocabulary by offering example sentences that show how a word is typically used. The dictionary helps with questions of spelling and usage as well. If you wonder whether a particular word is too informal for your writing, or if you have concerns that a word might be offensive, the dictionary will give you this information too.

You should be aware that Canadian usage and spelling may follow either British or American practice, but usually combine aspects of both. There are a number of Canadian usage manuals available today that help you to be consistent in your approach. It's also a good idea to make sure that the language option in your word processing program is set to *English* (*Canada*). However, when you are a professional engineer writing for American clients, you will want to use American spelling.

A thesaurus lists words that are closely related in meaning. It can help when you want to avoid repeating yourself, or when you are fumbling for a word that's on the tip of your tongue. Your word processor makes it easy to look up synonyms and antonyms. Be careful, though: you need to distinguish between **denotative** and **connotative** meanings. While a word's denotation is its primary or "dictionary" meaning, its connotations are any associations that it may suggest. They may not be as exact as denotations, but they are part of the impression a word conveys. If you examine a list of proposed synonyms, you will see that even words with similar meanings can have quite different connotations. For example, for the word *uncertain* your thesaurus may suggest the following: *unsure, vague, doubtful, hesitant, undecided, indecisive, ambiguous, ambivalent, unclear.* Imagine the different impressions you create in choosing one or the other of those words to complete this sentence: "He was _____ about the experiment's chance of success." To write effectively, you must remember that a reader may react to the suggestive meaning of a word as much as to its "dictionary" meaning.

Avoid jargon

All fields have their own terminology, or *occupational dialect.* It may be unfamiliar to outsiders, but it helps specialists explain things to each other. As a student, you are generally writing for experts, so there's no need to define standard technical terms or explain a methodology that's familiar to anyone with scientific training. In fact, by using the vocabulary of your discipline appropriately and correctly, you confirm that you are a serious student and a credible writer.

The trouble is that people sometimes try to use technical language to make themselves look more professional. Too often the result is not clarity but confusion, especially if the words aren't used correctly. The guideline is easy: use specialized terminology only when it's called for to explain something more precisely and efficiently. If plain language will do just as well, use it, especially if your reader is not an expert. In particular, avoid using scientific-sounding words in contexts that are not scientific:

orig. Consultation with managerial personnel furnished input for determining the viable parameters of the project. (14 words)

rev. The manager helped define the scope of the project. (9 words)

Be precise

As a scientist, you already know how to be exact. Carry this habit into your writing: always be as precise as you can. Avoid all-purpose adjectives like *major, significant,* and *relevant,* abstract general nouns like *situation* and *factor,* and vague verbs such as *involve, entail,* and *exist,* when you can be more specific:

> *orig.* Ensuring immediate access to emergency water supplies is a major element in effective seismic disaster management. (16 words)
>
> *rev.* After an earthquake, rescuers must first ensure access to emergency water supplies. (12 words)

Avoid generalizing with such qualifiers as *fairly, rather, somewhat,* and *quite.* Indeed, saying that something is *very* important carries less weight than saying simply that it is important. For example, compare these sentences:

> This is an important decision.
>
> This is a really important decision.

The shorter sentence has more impact. When you think that a word needs qualifying—and sometimes it will—first see if there's a more precise way of phrasing it. For example, the word *critical* conveys a greater degree of urgency than *important* and is more precise than something like *really important;* using *critical* will give a recommendation more weight:

> This is a critical decision.

If you are making qualitative judgments, you must be sure to back them up with specifics. In particular, watch how you use the word *relatively.* Its potential for misinterpretation is great unless you are actually considering two or more items. What does *relatively complex* mean anyway? Two other such weasel words are *basically* and *virtually.* Finally, watch how you use the qualifier *approximately.* Do not introduce an exact measure with this word:

> ✗ The receiver height was set at approximately 4.53 m.
>
> ✓ The receiver height was set at 4.53 m.

Be careful to avoid the mistake of qualifying absolutes. To say that something is *very unique* makes as little sense as describing something as *rather rectangular.* The following are other redundant descriptions to avoid. In each case, you can safely delete the underlined word.

in <u>close</u> proximity	revert <u>back</u> to
many <u>different</u> types	<u>future</u> plans
<u>sudden</u> crisis	<u>personal</u> opinion
this <u>particular</u> context	<u>successful</u> achievements
each <u>individual</u> participant	<u>true</u> facts

Avoid ambiguity

One good reason for having someone else read your final draft is to check for ambiguity—words or phrases with unintentional double meanings. For example, if you say "the *background* of the site was discussed," are you referring to the physical location or the context? Watch for words with multiple meanings, like *sound* and *solid.* Be particularly careful of the potentially confused meanings of *as* and *since* to refer to time or cause. You may find it safer to use unambiguous synonyms instead:

orig. Sediments will be trapped <u>as</u> runoff is detained in the storm water management basins.

rev. Sediments will be trapped <u>when</u> runoff is detained in the storm water management basins.

rev. Sediments will be trapped <u>because</u> runoff is detained in the storm water management basins.

Pronouns can cause trouble too, especially if your reader can't tell what word the pronoun is referring to:

orig. This seed requires water to germinate, and it must be warm.

Is it the seed or the water that must be warm? If there is any chance of ambiguity, restructure the sentence.

rev. These seeds require water to germinate, and they must be warm.

rev. These seeds require warm water to germinate.

One of the most commonly misinterpreted pronouns is the demonstrative *this*. In the example below, it's just not possible to tell whether it's the warning, the possibility, or the destruction that upsets taxpayers:

> *orig.* Experts warn that acid rain is destroying Canadian maple forests. This upsets most taxpayers.

> *rev.* The possibility that acid rain is destroying Canadian maple forests upsets most taxpayers.

Make sure there is just one word that the pronoun clearly refers to.

Avoid noun clusters

One recent trend for reducing sentence length is to use nouns as modifiers (as in the phrase *noun cluster*). In pairs, this practice is certainly useful: *kidney disease, hydrogen bomb, word processor, reaction time,* and *S.I. units* are all well-established noun pairs. But avoid making up your own combinations, like *resonator junction isolator,* where it is difficult for the reader to understand if you are talking about *junction isolators for resonators* or *isolators for resonator junctions.* Frequent use of extended series can produce monstrous pile-ups. Breaking up noun clusters may not result in fewer words, but it will make your writing easier to read:

> *orig.* computer services management group

> *rev.* group of computer services managers

> *rev.* group managing computer services

> *orig.* volunteer pollution investigation committee

> *rev.* volunteer committee to investigate pollution (*not* group investigating volunteer pollution!)

Be Concise

At one time or another, you will probably be tempted to pad your writing. Whatever the reason—because you need to write 2,000–3,000 words and have only enough to say for 1000, or just because you think length is strength and hope to get a better mark for the extra words—padding is a mistake. Readers suspect, quite reasonably, that you are only pretending to have something important to say.

Strong writing is always concise. It leaves out anything that does not serve some communicative or stylistic purpose. Concise writing conveys a better impression in both assignments and exams, for you never look as though you are babbling helplessly. Practise editing for economy by counting the words in your sentences and reducing that number whenever possible. The following guidelines can help.

Use adverbs and adjectives sparingly

Avoid the scattergun approach to adverbs and adjectives. Don't use combinations of modifiers unless you are sure they refine your meaning. One well-chosen word is always better than a series of synonyms:

> *orig.*   As well as being costly and financially extravagant, the venture is reckless and foolhardy. (14 words)
>
> *rev.*   The venture is both costly and foolhardy. (7 words)

Avoid chains of relative clauses

Sentences full of clauses beginning with *which*, *that*, or *who* are usually wordier than necessary. Try reducing some of those clauses to phrases or single words:

> *orig.*   The solutions <u>that</u> were discussed last night have practical applications, <u>which</u> will be easily understood by people who have no technical training. (22 words)
>
> *rev.*   The practical applications of the solutions discussed last night are easily understood by non-technical people. (15 words)

Reduce clauses to phrases or words

Independent clauses can often be reduced by subordination. Here are a few examples to imitate:

> *orig.*   The report was written in a clear and concise manner, and it was therefore immediately approved. (16 words)
>
> *rev.*   Written in a clear and concise manner, the report was immediately approved. (12 words)
>
> *rev.*   The clear, concise report was immediately approved. (7 words)
>
> *orig.*   The plan was of a radical nature and was a source of embarrassment to management. (15 words)
>
> *rev.*   The radical plan embarrassed management. (5 words)

Watch for ineffective or accidental repetition

Although your word processor will point out where you've inadvertently typed the same word twice, you'll need your eyes to catch most redundancies. This is another good reason to proofread carefully:

> *orig.* The terrain <u>slopes</u> to the south with <u>slopes</u> of up to 6.0%.
>
> *rev.* The terrain slopes to the south at angles of up to 6.0%.

Sometimes the repetition comes in a different part of speech. It still calls for a revision:

> *orig.* The architects will issue a certificate of <u>compliance</u> showing that construction <u>complies</u> with building codes. (15 words)
>
> *rev.* The architects will certify that construction complies with building codes. (10 words)

Eliminate hackneyed expressions and deadwood

Because they are so common, hackneyed phrases can quickly come to mind when you're writing. Unfortunately, readers find them stale and un-original. Unnecessary words are deadwood. To keep your writing vital, chop ruthlessly:

Wordy	Revised
at the same time as	while
due to the fact that	because
at the end of the day	finally, all in all
at this point in time	now
consensus of opinion	consensus
despite the fact that	although
in the near future	soon
when all is said and done	[omit]
in the eventuality that	if
in all likelihood	likely
it could be said that	possibly, maybe
in all probability	probably
last but not least	finally

Avoid "it is" and "there is" beginnings

Many readers object to sentences that start with *It is* or *There are* because such writing often seems inflated. But it may not always be possible to avoid beginning sentences this way. For one thing, such constructions help you avoid a personal focus:

orig.   We have reason to be optimistic.

rev.   There is reason for optimism.

If your sentence includes a clause beginning with *that*, however, you may want to do some editing:

orig.   There are several problems that need to be resolved. (9 words)

rev.   Several problems need to be resolved. (6 words)

Be on the lookout for sentences that can be tightened up considerably by rewriting:

orig.   It is the purpose of the project to pursue new software development. (12 words)

rev.   The project will develop new software. (6 words)

orig.   It is certain that inflation will increase. (7 words)

rev.   Inflation will certainly increase. (4 words)

Use vigorous verbs

Verbs like *have*, *do*, and *make* often introduce wordy phrases that you can reduce by choosing an appropriate strong verb instead:

Wordy	*Strong*
come to a conclusion	conclude
have a tendency to	tend to
do an analysis of	analyze
do research on	study, investigate
make a discovery	discover
make an effort to	try

Use verbs in the present tense rather than the future

Future verb forms add the auxiliary *will* to the infinitive to refer to an action that takes place later. You save words if you can write in the simple present tense instead:

> *orig.*   an action that will take place later (7 words)
>
> *rev.*   an action that takes place later (6 words)

Be Convincing

Developing a confident, readable style means learning some standard techniques and practising them until they become habit.

Choose active over passive verbs

Active sentences are usually livelier and shorter than passive ones:

> active: She presented the findings. (4 words)
>
> passive: The findings were presented by her. (6 words)

Moreover, passive constructions tend to produce awkward, convoluted sentences. Writers of bureaucratic documents are among the worst offenders:

> *orig.*   It had been decided that the utilization of small rivers in the province for purposes of hydroelectric power generation should be studied by the department and that a report to the deputy minister should be made by the director as soon as possible. (43 words)

The passive verbs in this mouthful detract from the issue and leave the sentence looking as though there's something to conceal. If a passive construction buries the "doer" of an action in a phrase beginning with "by," you should rewrite to save words and emphasize the subject:

> *rev.*   Once the department has investigated using small rivers to generate hydroelectric power in the province, the director will immediately report to the deputy minister. (24 words)

To focus on results rather than on the people producing those results, scientific writing often relies on passive constructions—quite appropriately. In fact, four situations actually call for the passive rather than the active wording:

1. When you want to emphasize the results of actions rather than the person performing the action or achieving the results. This is generally the situation in lab reports:

   > As a result, the concentration of radon <u>was reduced</u> by 40 per cent.

2. When the subject is the passive recipient of some action:

   > The university <u>was founded</u> in 1959.

3. When you want to avoid an awkward shift from one subject to another in a sentence or paragraph:

   > The sensor monitors light and <u>is checked</u> once a day.

4. When you want to avoid assigning responsibility or blame:

   > Several errors <u>were made</u> in the calculations.

When passive verbs produce wordy or convoluted constructions, however, be sure to rewrite the sentence:

orig. If the fan <u>is located</u> in a remote space, noise <u>will be minimized</u>, and the fan will thus <u>be able to be operated</u> throughout the night. (26 words)

rev. <u>Minimizing</u> noise by <u>locating</u> the fan away from the bedrooms will permit <u>its operation</u> throughout the night. (17 words)

If there are concise alternatives to passives, use them:

orig. Software reengineering <u>is concerned with</u> the redesign or reconstruction of a software system. (13 words)

rev. Software reengineering <u>involves</u> the redesign or reconstruction of a software system. (11 words)

Use strong subjects

In sentences that feature people and actions, it makes sense not to bury these in complicated phrases. When you have a choice, make the subject of your sentences concrete, not abstract:

orig.   At the beginning of the term, instructions for handling hazardous materials were given to each student. (16 words)

rev.   At the beginning of the term, students were given instructions for handling hazardous materials. (14 words)

In most cases this principle calls for an *active* instead of a *passive* verb:

orig.   When the function key is pressed by the operator, the RX screen is activated. (13 words)

rev.   The operator presses the function key to activate the RX screen. (11 words)

If you are writing instructions, you must address them to the person who will be following them. You accomplish this by using the *imperative*—a one-word verb with *you* understood as the subject. Note the difference below:

orig.   To record the results, a three-step process must be followed. (10 words)

rev.   Follow a three-step process to record the results. (8 words)

Avoid hesitancy

It's normal for students to worry that they lack the knowledge or experience to give a subject the development that it deserves. Don't apologize for your inexperience, however. State your information plainly and without qualifications. Use words like *seem* and *appear* only in situations of genuine doubt, not because you are afraid you have not done enough research:

orig.   The tank appears to be located so as to be fairly accessible for periodic cleanout. (15 words)

rev.   The tank's location makes it accessible for periodic cleanout by a professional contractor. (13 words)

Be Consistent

The demand for readability also calls for consistency. There should never be surprises for your reader. If you are constantly changing point of view or attitude or tone, your reader can't believe that you are in control of your ideas. After all, it's hard to trust someone who keeps changing the rules. Watch for and avoid situations where the wording seems to shift.

Keep a consistently appropriate tone

In Chapter 1, you learned about choosing a tone that matches the context of your writing (see pp. 9–10). If you think of this in terms of clothing, it means selecting something to wear that makes you look good (not sloppy), yet doesn't make you feel uncomfortably overdressed. The corresponding tone and register lie generally between standard and formal. What's even more important, whether you are writing a cover letter or a dissertation, is that you not mix levels from sentence to sentence, or even within a sentence. Lady Gaga may be able to wear a three-foot headdress while running errands, but you should not. Consider this example from a cover letter for a job application:

> *orig.* As a project team member, I interface with the project leader, project members, but pretty rarely with the end consumer. (20 words)

Apart from faulty parallel structure (see pp. 200–201), this sentence alternates Standard English, jargon, and slang. The solution is to revise to make the language consistently standard:

> *rev.* On projects, I work with the project leader and team members, rarely with the consumer. (15 words)

Another typical inconsistency is a shift in tone from impersonal to personal, which often coincides with a shift from formal to informal:

> *orig.* The report discusses which recommendations should (in our opinion) get priority attention. A general plan of attack for making it past this phase will be included. (26 words)

> *rev.* The report discusses the recommendations that deserve primary attention. It includes a master plan and schedule. (16 words)

Generally the solution is to keep to the middle ground—not too formal, not too casual.

Shifting from one pronoun to another will also affect the consistency and formality of your writing. Where "you" is often too personal and too casual (except in personal communications and instructions), "one" is unnaturally formal. In your writing, then, avoid generalizing with either of these choices, and be especially careful not to combine the two approaches in the same document:

> *orig.* One's resumé must look professional, and you should always include a cover letter.

> *rev.* A resumé must look professional, and it should be accompanied by a cover letter.

Keep a consistent point of view

Where writers might once have referred to themselves, awkwardly, in the third person (as "the investigators" or "the authors"), scientific writing now generally permits "we" and a first-person point of view, especially when it makes sentences more readable. The level of formality is essentially your choice, but remember that shifts from personal to impersonal are awkward in writing:

> *orig.* We note the deformation of the surface, and it is recommended that cryogenic treatment be attempted. (16 words)

> *rev.* We note the deformation of the surface and recommend cryogenic treatment. (11 words)

Notice that the revised sentence not only offers a consistent point of view but also features an active rather than a passive verb and is shorter, thus stronger.

Be Objective

Your writing must be free of biases and subjective opinions. Your reader will be more likely to accept your findings if you follow these suggestions:

- Avoid unsubstantiated judgments. Be sure that any suggestions you make or conclusions you reach follow from the information you have provided. Never imply anything that you cannot prove. If your findings aren't foolproof, add credibility by showing where the uncertainty lies.

- Avoid subjective language. Words such as *terrible* or *excellent* detract from the objective tone you want. Rather than saying "quarterly figures show an incredible increase," give the exact percentage of the increase and let the facts speak for themselves.

The overall tone to aim for depends on the circumstances—and particularly on the intended reader. If you are writing a short, informal report to someone you see often, familiar terms such as *I* or *you* do work well. On the other hand, many formal reports try to avoid the subjectivity of personal pronouns. The result is often a cumbersome load of passive constructions (see pp. 169–170). If you are writing on behalf of a group or a team, you can use the pronoun *we*, which avoids awkward passive constructions. If you are writing as an individual, you don't have that option. In such a case, try to recast the sentence to keep an active verb while avoiding *I*:

✗ The purchasing system has been found to increase the duplication of forms. (12 words)

✗ I found that the purchasing system increases the duplication of forms. (11 words)

✓ The purchasing system increases the duplication of forms. (8 words)

If you can't make this sort of revision, you would do better to write the occasional *I* than to use convoluted passive constructions that strangle your meaning or substitutes like *the writer* or *the researcher* that make your writing seem old-fashioned.

Using inclusive language

For the sake of credibility, avoid any suggestion of bias in the language you use, whether written or spoken. Just as you consider your reader carefully when deciding on the appropriate tone to use, you also have to be careful to use neutral language.

The potential for bias is far-reaching, involving gender, race, culture, age, disability, occupation, religion, and socio-economic status. Although society hasn't come up with ideal solutions in every case, developing an awareness of sensitive issues helps you to avoid using non-inclusive language.

Gender

At one time, it was common to use *he* as a generic singular pronoun. Indeed you will still encounter this usage in books published before 1975:

> If an employee discovers a way to cut costs, he receives a bonus.

Informally, we've solved the problem by replacing *he* with *they*, but this solution is not universally accepted in formal writing. Rather than use an awkward combination of singular and plural, try these options for avoiding the problem:

- Pluralize the word and the pronouns that refer to it:

  > If employees discover a way to cut costs, they receive a bonus.

- Use the passive voice:

  > A bonus is given to an employee who discovers a way to cut costs.

- Restructure the sentence:

  > An employee who discovers a way to cut costs receives a bonus.

- Use *he or she* as a last resort. (It becomes cumbersome and annoying if used repeatedly.)

- Avoid the abbreviated forms *he/she* and *s/he* except in notes.

Race and culture

Words attached to racial or cultural identity sometimes carry negative connotations, for example the term *Negro*. The search for neutral language has produced alternatives such as *black, brown,* or *person of colour*, but these have not been universally accepted. There are similar problems with the term *Indian*, with alternatives such as *Aboriginal, Native, Indigenous,* and *First Nations* each having its share of critics. The best solution is often to find out what the group in question prefers. In scientific and technical writing, you are not likely to be dealing with the same issues of race, religion, or culture as in a sociology course. Still, it's important to remember that any reader may be highly sensitized to bias. It makes good sense to seek and use the most objective language you can.

There are areas beyond gender and ethnicity where the effort to develop and use neutral language has made an impact. For instance, we refer not to *old people* but to *seniors*, and to someone as having *special needs* rather than being *handicapped*. Whatever the situation, be sensitive to the power of the words you use and take the time to search for language that is bias-free.

Make Important Ideas Stand Out

Experienced writers know how to construct sentences to emphasize certain points. The following are some of their techniques.

Use concrete details

Concrete details are easier to understand than abstract ideas—and more meaningful. If you are writing about abstract concepts for readers who are not experts in your field, be sure to provide specific examples and illustrations:

> *orig.* The following are scientists who are dealing with problems associated with stochastic process: physicists, meteorologists, economists, and so on.

> *rev.* The physicist measuring frequencies in the lab, the meteorologist forecasting rain, the economist verifying price fluctuations—all of these scientists are tackling problems related to stochastic process.

See how a few specific details can bring the facts to life? Adding concrete details and examples is another way to improve the readability of your writing.

Place key words in strategic positions

The positions of emphasis in a sentence are the beginning and, above all, the end. If you want to make a point convincingly, don't bury it in the middle of the sentence. Feature it later:

> *orig.* The strongest part of the committee's report is its recommendations, not its findings.

> *rev.* The strongest part of the committee's report is not its findings but its recommendations.

Subordinate minor ideas

It is easy to connect incidents with a string of *and*s, as if everything were of equal importance:

> We performed the experiment, and we analyzed the results, and we made firm recommendations.

But you'll find it just as natural to *subordinate*—that is, to make one part of a sentence less important grammatically in order to emphasize another point:

> Once we performed the experiment and analyzed the results, we made firm recommendations.

Major ideas stand out more and connections become clearer when minor ideas are subordinated:

orig. Fumes from insulation materials can build up along with carbon dioxide, mould, bacteria, and dust, and the result may be "sick building syndrome."

rev. When fumes from insulation materials build up along with carbon dioxide, mould, bacteria, and dust, the result may be "sick building syndrome."

Make your most important idea the subject of the main clause. If you put this main clause at the end (see *periodic sentence* in the Glossary), it will carry more weight:

orig. "Sick building syndrome" may result when fumes from insulation materials build up.

rev. When fumes from insulation materials build up, "sick building syndrome" may result.

Vary sentence structure

As with anything else, variety adds life to writing. One way of adding variety, as you have seen, is to change the sentence order. Most sentences follow a standard pattern of **subject + verb + object (+ modifiers)**:

> The class president circulated a petition last week.
>   subject   verb   object   modifier

Placing a modifier at the beginning rather than in its natural place at the end of the sentence calls attention to it:

orig. Unsolicited commercial e-mail, known as *spam*, is little more than a minor irritant <u>for most Internet users</u>.

rev. <u>For most Internet users</u>, unsolicited commercial e-mail, known as *spam*, is little more than a minor irritant.

A full inversion of a sentence creates the greatest emphasis because it is so unusual. Save it for those times when you want to make a very strong point:

orig. Astronomers <u>did not learn</u> of the existence of galaxies until the 1920s.

rev. <u>Not</u> until the 1920s <u>did</u> astronomers <u>learn</u> of the existence of galaxies.

Use comparison and contrast

Just as a jeweller will highlight a diamond by displaying it against dark velvet, so too can you highlight an idea by placing it against a contrasting background:

orig. A communications engineer conceives the overall design of a communications system.

rev. Unlike the electronics engineer, whose job it is to design equipment, the communications engineer conceives the system itself.

Using parallel phrasing increases the effect of the contrast:

Dioxin is <u>not</u> a name for one chemical <u>but</u> a general term for a group of chemicals—all of them toxic.

Parallel comparatives help you emphasize the increase or decrease of something in proportion to something else. The result is both economical and readable:

orig. The water contained in an aquifer increases proportionally to the number of pores in the aquifer, and its circulation speed increases as the pore size increases. (26 words)

rev. <u>The more</u> pores in an aquifer, <u>the more</u> water it contains; <u>the bigger</u> the pores, <u>the faster</u> the water circulates. (20 words)

Use correlative constructions

Correlatives such as *both . . . and* or *not only . . . but (also)* can be used to emphasize combinations as well, as long as you keep both sides balanced and parallel (see pp. 200–201). Note the options for increased and reduced sentence length:

orig. The model achieves the highest standards of energy efficiency, <u>and</u> it is cost effective <u>as well</u>. (16 words)

rev. <u>Not only</u> does the model meet the highest standards of energy efficiency, <u>but</u> it is cost effective <u>as well</u>. (19 words)

rev. The model is <u>both</u> energy efficient <u>and</u> cost effective. (9 words)

Vary sentence length

A short sentence can add impact to an important point, especially when it comes after a series of longer sentences. This technique can be particularly useful for conclusions. Don't overdo it, though. A string of long sentences may be monotonous, but a string of short ones can make your writing sound amateurish.

Still, scientific writing is likely to have more long sentences than short ones. Since short sentences are easier on readers, try breaking up clusters of lengthy ones. Check any sentence of over 20 words or so to see if it will benefit from being split. Use the readability indicator on your word processor to help you keep the average length of your sentences between 15 and 18 words. Never hesitate to tighten up a loose sentence either.

Use your ears

Your ears are probably your best guides. Make good use of them. Before producing a final copy of any piece of writing, read your draft out loud in a clear voice. The difference between cumbersome and fluent passages will be unmistakable.

Write Before You Revise

No one expects you to put all this advice into practice as soon as you sit down to write. The best time to edit is during the final stages of the writing process, when you are ready to look critically at what you have already written. Most writers find it easier to plan and prepare several drafts before starting the detailed task of revising and editing.

Once you are ready to edit, look carefully for ways to make your sentences shorter and more manageable. Note how the second of the following paragraphs makes its point in fewer words than the first.

orig. The benefits of the project will include an understanding of the groundwater resources of the areas, their location, magnitude, and recharge characteristics. This understanding will be refined in the immediate well field area to produce a definition of the capture zones that will provide the basis for delineating well head protection areas around each well or well field. Groundwater protection measures in these areas will therefore be recommended or required by policies and programs currently being developed.

The three sentences that make up this passage contain 77 words. The average sentence is 25.6 words long, and 50 per cent of the verbs are in the passive voice. This text is hard to read.

rev. The project identifies the groundwater resources of the area, as well as their location, magnitude, and recharge characteristics. These data help define the capture zones in the immediate well field area and delineate protection areas around each well head or field. Future policies and programs will recommend or require groundwater protection measures in these areas.

The number of words has been reduced to 55 from 77, leaving an average sentence length of 18.3 words with only 20 per cent passive verbs. The revision is much easier to read than the original.

The best advice is to take advantage of your situation as a student to try things out, to discover what your readers like or dislike, and to practise editing with the goal of reducing the average number of words in your sentences. By working at the readability of your writing, you will be well prepared for a career where your writing skills go hand in hand with your success.

Chapter Checklist

- ☐ Edit sensitively and carefully.
- ☐ Reduce sentence length.
- ☐ Choose active verbs and concrete, plain nouns.
- ☐ Be positive, and sound natural.
- ☐ Be consistent in tone, point of view, and diction.

Common Errors in Grammar and Usage

Introduction

This chapter surveys those areas where students most often make mistakes. It will help you watch for weaknesses as you edit your work. Once you get into the habit of checking, it won't be long before you avoid potential problems as you write.

The grammatical terms used here are the most basic and familiar ones; if you need some review, see Chapter 16 or the Glossary. For a thorough treatment of grammar and usage, consult a guide such as *The Canadian Writer's Handbook: Essentials Edition* (Toronto: Oxford University Press, 2012). Take advantage of your school's writing centre to benefit from personal attention.

If English is not your first language, a section at the end of this chapter focuses on the most typical errors of non-native speakers (see pp. 201–213). As you work to increase your proficiency, keep track of mistakes you typically make and focus on eliminating them as you edit. It is also a good idea to have your writing reviewed by a colleague whose first language is English. This person can point out idiomatic and other errors that you may have missed.

Problems With Sentence Unity

Sentence fragments

To be complete, a sentence must have both a subject and a predicate in an independent clause. If not, it's a fragment. There are times in informal writing when it is acceptable to use a sentence fragment in order to give emphasis to a point, as in

✓ Will the municipality reduce property taxes? <u>Not likely</u>.

The sentence fragment *Not likely* is deliberate. Because the writer was speaking directly to an audience, it is an appropriate shortening of *It is not likely that the municipality will reduce property taxes.* Unintentional sentence fragments, on the other hand, usually seem incomplete rather than shortened:

✗ One application is the algorithm for a chess program. <u>Based on an actual game situation.</u>

The last "sentence" is incomplete because it has neither a subject nor a verb. (Remember that a participle such as *based* is a verbal, or "part-verb," not a verb.) The fragment can be made into a complete sentence by adding a subject and a verb:

✓ It is based on an actual game situation.

More economically, you could join the fragment to the preceding sentence:

✓ One application is the algorithm for a chess program based on an actual game situation.

Be particularly careful not to separate dependent clauses from the previous sentence. Watch for such subordinators as *whereas, while,* and *because.*

✗ Some companies are exemplary corporate citizens. <u>Whereas others are interested only in making profits.</u>

One good test for a complete sentence is to make sure that it fits naturally into the following space: "It is true that _____." If the result sounds wrong— as it would with the sentence fragment beginning *whereas* above—it makes sense to revise it:

✓ Some companies are exemplary corporate citizens, whereas others are interested only in making profits.

Run-on sentences

Many people consider a run-on sentence one that continues beyond the point where it should have stopped:

✗   Mosquitoes and blackflies are annoying, but they don't stop tourists from coming to spend their holidays in Canada, and such is the case in Ontario's northland.

This example reveals a problem of over-coordination. The sentence could be improved by removing the word *and* and replacing the comma after Canada with a semicolon or period.

The grammatical problem called a run-on sentence occurs when two independent clauses are jammed together without any punctuation at all. (An independent clause is always a complete sentence.) Two independent clauses should not run together, as they seem to do in the following example:

✗   Glaciers are retreating so fast scientists are concerned about global water levels.

✓   Glaciers are retreating so fast that scientists are concerned about global water levels.

Here the problem is corrected by adding the word *that*, which makes it clear that the second clause (*scientists are concerned about global water levels*) is a dependent, or subordinate, clause. The problem cannot be corrected by simply adding a comma, as the writer of the next example has done:

✗   In physics there is matter and antimatter, everything must be balanced.

This error is known as a comma splice. There are three ways of correcting it:

1. by putting a period after *antimatter* and starting a new sentence:

   ✓   . . . and antimatter. Everything . . .

2. by replacing the comma with a semicolon:

   ✓   . . . and antimatter; everything . . .

3. by making one of the independent clauses subordinate to the other, so that it can't stand by itself:

✓ In physics, <u>because</u> there is matter and antimatter, everything must be balanced.

The comma splice is occasionally forgivable when clauses are very short and arranged in tight sequence:

orig.   We noted the contrast: the flow geometry was simple, the flow pattern was complex.

However, it's still better to revise such a sentence:

✓ We noted the contrast: the flow geometry was simple, and the flow pattern was complex.

Contrary to what many people think, words such as *however, therefore,* and *thus* should not be used with a comma to join independent clauses:

✗ The coefficient used for the calculation is still correct, <u>therefore</u> no new modelling has been done.

Correct this mistake by beginning a new sentence after *correct* or by replacing the comma with a semicolon:

✓ The coefficient used for the calculation is still correct; <u>therefore</u>, no new modelling has been done.

The only words that can be used without a semicolon to join independent clauses are the coordinating conjunctions—*and, or, nor, but, for, yet,* and *so.* (Subordinating conjunctions such as *if, because, since, while, when, where, after, before,* and *until* reduce the construction to a dependent clause.)

✓ The coefficient used for the calculation is still correct, <u>so</u> no new modelling has been done.

Faulty predication

When the subject of a sentence is not grammatically connected to what follows (the predicate), the result is called *faulty predication*:

✗ The <u>reason</u> they chose this design is <u>because</u> it is well suited to indoor environments.

The problem with this sentence is that *because* means essentially the same thing as *the reason (that)*. The subject needs a noun clause to complete it:

✓ The <u>reason</u> they chose this design is <u>that</u> it is well suited to indoor environments.

Another solution is to rephrase the sentence:

✓ They <u>chose</u> the design <u>because</u> it is well suited to indoor environments.

Faulty predication also occurs with *is when* and *is where* constructions:

✗ The best time for the tests <u>is when</u> the subjects are well rested.

Again, you can correct this error in one of two ways:

1. Follow the *is* with a noun phrase to complete the sentence:

   ✓ The best time for the tests <u>is the morning, when</u> the subjects are well rested.

2. Change the sentence:

   ✓ Run the tests in the morning, when the subjects are well rested.

Problems With Subject-Verb Agreement

Identifying the subject

Formal writing always requires the verb to agree in number with its subject. In other words, singular subjects call for singular verbs; plural subjects call for plural verbs. Sometimes, however, when the subject does not come at the beginning of the sentence, or when it is separated from the verb by other information, you may inadvertently use a verb form that does not agree, as in the example that follows:

✗ Our design <u>work</u>, including earthwork calculations and cost estimates, <u>are based</u> on these figures.

The subject here is *work*, not *calculations and estimates*; therefore, the verb should be singular:

> ✓ Our design <u>work</u>, including earthwork calculations and cost estimates, <u>is based</u> on these figures.

Either, neither, each

The indefinite pronouns *either*, *neither*, and *each* always take singular verbs:

> ✗ Neither of the models <u>are</u> flawless.
>
> ✓ Neither of the models <u>is</u> flawless.
>
> ✓ Each of them <u>has</u> flaws.

Compound subjects

When you use *or, either . . . or,* or *neither . . . nor* to create a compound subject, the verb is expected to agree with the last item in the subject:

> ✓ Neither the TA nor <u>her students were</u> able to solve the equation.
>
> ✓ Either the students or <u>the TA was</u> misinformed.

You may find that it sounds awkward to use a singular verb when a singular item follows a plural item:

> *orig.*   Neither the transistors <u>nor the circuit was</u> modified.

In such instances, it's worth rephrasing the sentence:

> *rev.*   Neither the circuit nor <u>the transistors were</u> modified.

Unlike the word *and*, which creates a compound subject and therefore takes a plural verb, the phrases *as well as* and *in addition to* do not themselves make a subject plural. The verb agrees with the original subject:

> ✓ <u>Noise, interference, and channel distortion corrupt</u> information transmissions.
>
> ✓ <u>Noise</u>, as well as interference and channel distortion, <u>corrupts</u> information transmissions.

When the main verb is a form of *be* (*is*, *are*, *was*, or *were*), you often have a choice of sentence order. Note the subject-verb agreements below:

✓ My <u>area of research</u> <u>is</u> biomechanics and motor control.

✓ <u>Biomechanics and motor control</u> <u>are</u> what I study.

Remember that you can always revise your sentences to make them stronger.

Subject-verb agreement is often ignored in spoken English (which is much less formal than writing, remember). Orally, sentences beginning with *there*, for example, tend to be followed by the singular verb, even when the subject is a plural. Avoid such casual structures in writing:

✗ There's a number of <u>problems</u> to be resolved.

✓ There <u>are</u> a number of <u>problems</u> to be resolved.

✓ A number of <u>problems</u> <u>need</u> to be resolved.

Subject-verb agreement errors sometimes result from a misunderstanding of the true subject in phrases containing *of*. In the example above, *a number of* is grammatically equivalent to *many*, so it is appropriate to use a plural verb. But many similar constructions are singular, not plural, because the true subject comes after the definite or indefinite article (*the* or *a*), not *of*. When you use the following, be sure to use a singular verb:

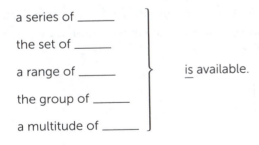

a series of _____

the set of _____

a range of _____   <u>is</u> available.

the group of _____

a multitude of _____

Collective nouns

A collective noun is a singular noun that represents a number of members (examples include words such as *family*, *class*, *group*, and *team*). If the noun refers to the members as one unit, it takes a singular verb:

✓ The <u>class</u> <u>is</u> studying communication systems.

If, in the context of the sentence, the noun refers to the members as individuals, the verb becomes plural:

✓ The <u>class</u> <u>are</u> submitting their projects on Friday.

It's advisable to rewrite such a sentence to avoid the awkward sound of the singular and plural together:

✗ The <u>class</u> <u>is</u> submitting <u>their</u> projects on Friday.

✓ The <u>class</u> <u>is</u> submitting <u>the</u> projects on Friday.

✓ The <u>students</u> <u>are</u> submitting <u>their</u> projects on Friday.

Quantities

A number of nouns and pronouns are used to measure quantities that can be either enumerated or portioned out (see pp. 201–205 for a discussion of countable and uncountable nouns). You must be especially careful to distinguish between singulars and plurals so that your choice of verb will be appropriate:

✓ Most of the <u>noise</u> <u>was</u> eliminated.

✓ Most of the <u>noises</u> <u>were</u> accounted for.

With the words that follow, the rule is to use a plural verb if you are measuring a number of individual items (*microchips, electrodes, test tubes*) and a singular verb if you are measuring a portion of something that cannot be counted (*equipment, software, water*):

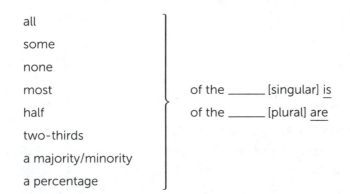

all	
some	
none	
most	of the _____ [singular] is
half	of the _____ [plural] are
two-thirds	
a majority/minority	
a percentage	

The following sentences show the difference:

✓   None of the <u>experiment</u> <u>was</u> videotaped.

✓   None of the <u>experiments</u> <u>were</u> videotaped.

Unusual plurals

A number of nouns cause real trouble for writers because they do not follow the traditional pattern of forming plurals by adding an *s*.

singular	*plural*
datum	data
criterion	criteria
phenomenon	phenomena
stratum	strata

Watch for these words in your writing, and be sure to use them correctly:

✗   The <u>criteria</u> <u>was</u> met.

✓   One criterion was met.

✓   <u>One</u> of the criteria <u>was</u> met.

✓   <u>All the criteria</u> <u>were</u> met.

In addition to these, there are a number of words that form plurals by adding *s* only in certain instances and have alternative plural forms in other contexts. Consult a dictionary before using words like *medium* and *antenna* in the plural.

Finally, be aware that in scientific writing, *data* is considered plural. In non-scientific contexts, however, *data* is accepted as a singular synonym for *information*.

Titles

The name of a business or organization is always treated as a singular noun, even if it contains plural words. The same is true of book titles. Use a singular verb with these:

✓   <u>Essentials of Optics</u> <u>is</u> an excellent textbook.

✓   <u>Goodman & Goodman</u> <u>is</u> handling the legal dispute.

Tense Problems

When you are speaking, your tenses usually come automatically, but it's easy to run into difficulty when writing. A few general rules can help you avoid problems.

The past perfect

If you have a reference point in the past and you want to mention something that happened *prior to* that time, use the *past perfect* (*had* followed by the past participle). The time sequence will *not* be clear if you use the simple past for both:

✗ The second study <u>revealed</u> that the earlier results <u>were misinterpreted</u>.

✓ The second study <u>revealed</u> that the earlier results <u>had been misinterpreted</u>.

Similarly, when you are reporting what someone said in the past—that is, when you are using *past indirect discourse*—use the past perfect form to distinguish what's happening at the time from what happened prior to that time:

✗ The CEO said that the project <u>was approved</u>.

✓ The CEO said that the project <u>had been approved</u>.

If–then conditions

When you are describing regularly occurring consequences, use the present tense in both the condition (*if*) clause and the consequence (*then*) clause:

✓ If the temperature ever <u>drops</u> below −6°, the liquid <u>freezes</u>.

✓ Whenever the temperature <u>drops</u> below −6°, the liquid <u>freezes</u>.

When you are predicting a specific future consequence, use the present tense in the *if* clause and the future in the *then* clause:

✓ If the temperature <u>drops</u> below −6° tonight, the liquid <u>will freeze</u>.

When the situation is hypothetical, it is conventional—especially in formal writing—to use the *subjunctive* form in the condition clause and *would* + the base verb in the consequence clause:

✓   The experiment <u>would fail</u> if the solution <u>precipitated</u>.

Note that the subjunctive form is exactly the same as the past tense. The subjunctive form of the verb *be* is always *were*. Another way of expressing this subjunctive is to use *were to* + the base verb:

✓   If the solution <u>were to precipitate</u>, the experiment <u>would fail</u>.

When you are describing a hypothetical instance in the past, use the *past subjunctive* (it has the same form as the past perfect) in the *if* clause and *would have* + the past participle for the consequence. It is an error to use *would have* in both clauses:

✗   If the solution <u>would have precipitated</u>, the experiment <u>would have failed</u>.

✓   If the solution <u>had precipitated</u>, the experiment <u>would have failed</u>.

Writing about technology

When you are describing a situation with a historical context, use the past tense:

✓   In 1993, Claude Berron <u>introduced</u> a new class of channel encoders, which he <u>named</u> turbo codes.

To discuss applications that are timeless (sometimes called "scientific truths") use the present tense:

✓   Turbo codes <u>achieve</u> high rates with low interference.

To establish a relationship between past and present in writing, use the present perfect form (*have* + the past participle):

✓   Recently, LPDC codes <u>have outperformed</u> turbo codes.

Do not hesitate to combine past and present tenses in the same paragraph as long as the contexts are appropriate. Never shift without good reason:

> ✓ Tests <u>were performed</u> on 14 November, and results <u>have been ana-</u><u>lyzed</u>. The company now <u>recommends</u> further study.

Maintain the present tense for all reports of current or timeless actions, processes, or states.

Pronoun Problems

Pronoun reference

The noun a pronoun refers to is called a *referent* or *antecedent*. There needs to be a clear antecedent for every pronoun. If the referent doesn't appear in the same sentence, it must appear in the preceding one:

> ✗ <u>Groundwater</u> specialists have developed a variety of strategies to de-contaminate <u>it</u>.

Even though *groundwater* appears in the sentence, it is used here as a modifier, not a noun. Therefore, it cannot serve as referent or antecedent for the pronoun *it*. Either replace it or rephrase your sentence:

> ✓ <u>Specialists</u> have developed a variety of strategies to decontaminate <u>groundwater</u>.

When a sentence contains more than one noun, make sure there is no ambiguity about the antecedent:

> ✗ The public wants increased <u>environmental responsibility</u> along with <u>lower taxation</u>, but the government does not favour <u>it</u>.

What does the pronoun *it* refer to: *responsibility, taxes,* or even *the public*?

> ✓ The public wants increased <u>environmental responsibility</u> along with <u>lower taxes</u>, but the government does not advocate <u>spending</u> <u>increases</u>.

Another problem with pronouns relates to the same singular-plural agreement issues that occur with subjects and verbs (see pp. 185–188). Singular pronouns have singular nouns as antecedents; plural pronouns refer to plural nouns:

✗ A spokesperson for GVA <u>Transit</u> said that <u>their</u> ridership had doubled in the past year.

✓ A spokesperson for GVA <u>Transit</u> said that ridership had doubled in the past year.

In speaking, it is common to use *they* and *their* to refer to a singular noun—particularly when the noun is a person of unknown gender—but this practice does not yet extend to formal writing. Rewrite the sentence to avoid the problem, either by pluralizing everything or dropping the pronouns:

✗ <u>Every</u> co-op <u>student</u> must submit <u>their</u> work report by the beginning of <u>their</u> next school term.

✓ <u>All</u> co-op <u>students</u> must submit <u>their</u> work <u>reports</u> by the beginning of <u>their</u> next school term.

✓ <u>Every</u> co-op <u>student</u> must submit <u>a</u> work report by the beginning of <u>the</u> next school term.

Using "it," "which," and "this"

Using *it*, *which*, and *this* without a clear referent can lead to confusion:

✗ Although the directors wanted to meet in January, <u>it</u> [<u>this</u>] didn't take place until March.

✓ Although the directors wanted to meet in January, <u>the conference</u> didn't take place until March.

When you use *which*, make sure that it refers to its noun antecedent and not to a general idea:

✗ The directors wanted to meet in January, <u>which</u> didn't take place until March.

✓ The directors wanted <u>a winter meeting</u>, <u>which</u> didn't take place until March.

If you don't want your sentences to confuse readers, make sure that your pronoun clearly refers to a specific, identifiable referent.

Using "one"

People sometimes use the pronoun *one* to avoid *I* in formal writing. Although common in Britain, such a reference is too formal for a North American audience:

> orig. If <u>one</u> were to apply for the grant, <u>one</u> would find <u>oneself</u> engulfed in endless red tape.

While there is nothing grammatically incorrect in this example, it sounds pretentious. The best alternative is to recast the sentence with a plural subject:

> rev. <u>Researchers</u> applying for the grant could find <u>themselves</u> engulfed in endless red tape.

If you avoid using *one*, you automatically avoid making the error of shifting point of view (mixing the third-person *one* with the second-person *you)*:

✗ When <u>one</u> uses a bank machine, <u>you</u> are using a CNC system.

✓ A <u>person</u> using a bank machine is using a CNC system.

Using "me" and other object pronouns

Remembering that it is wrong to say "My supervisor and me were invited," many people use the subject form (*I*) of the pronoun when the object form (*me*) is correct:

✗ The committee invited Damian and <u>I</u> to present our findings.

✓ The committee invited Damian and <u>me</u> to present our findings.

The verb *invited* requires an object; *me* is the objective case. A good way to tell which form is correct is to ask yourself how the sentence would sound with only the pronoun. You will know by ear that the subject form—"The committee invited *I*"—is inappropriate.

Knowing that *I* should be avoided in the previous example, some people prefer to substitute *myself*, but this usage is equally ungrammatical. *Myself, yourself, ourselves*, and so on, are reflexive pronouns used when their referent has already appeared, usually as the subject, in the sentence:

✓ She wrote herself a note to remind herself to return the book.

Avoid using reflexive pronouns as substitutes for object forms:

✗ The final exam schedule causes problems for myself.

✓ The final exam schedule causes problems for me.

Problems also arise with prepositions, which should be followed by a noun or pronoun in the objective case:

✗ Between you and I, this result doesn't make sense.

✓ Between you and me, this result doesn't make sense.

There are times, however, when the correct case can sound stiff or awkward:

orig. The reporter wanted to know to whom the award had been given.

Rather than keep to a correct but awkward form, feel free to reword the sentence:

rev. The reporter wanted to know who had received the award.

Exceptions for pronouns following prepositions

The rule that a pronoun following a preposition takes the objective case has exceptions. When the preposition is followed by a clause, the pronoun should take the case required by its position in the clause:

✗ The students were curious about whom would be elected.

Although the pronoun follows the preposition *about*, it is also the subject of the verb *would be elected* and therefore requires the subjective case:

✓ The students were curious about who would be elected.

Similarly, when a gerund (an *-ing* word that acts partly as a noun and partly as a verb) is the subject of a clause or phrase, the word that modifies it takes the possessive form:

> ✗   Our drafting instructor objected to <u>us</u> asking for an extension.

> ✓   Our drafting instructor objected to <u>our</u> asking for an extension.

> ✓   The drafting instructor objected to the <u>students'</u> asking for an extension.

Problems With Modifiers

Adjectives modify nouns; adverbs modify verbs, adjectives, and other adverbs. Never use an adjective to modify a verb:

> ✗   He played <u>good</u>. (*adjective with verb*)

> ✓   He played <u>well</u>. (*adverb modifying verb*)

> ✓   He played <u>very well</u>. (*adverb modifying adverb*)

> ✓   He had a <u>fine style</u>. (*adjective modifying noun*)

> ✓   He had a <u>very fine style</u>. (*adverb modifying adjective*)

The examples above feature one-word adjectives and adverbs. Many combinations can act in similar ways—as adjectival and adverbial modifiers. The following represent some of the most common ones:

> She is a student <u>in residence</u>. (*adjectival*)

> The students live <u>in residence</u>. (*adverbial*)

> Here is a place <u>to begin</u>. (*adjectival*)

> <u>To begin</u>, open the book. (*adverbial*)

> The TA <u>preparing the slides</u> is new. (*adjectival*)

> The TA discovered an error <u>while preparing the slides</u>. (*adverbial*)

> Everyone understood the material <u>prepared by the TA</u>. (*adjectival*)

Being familiar with these types of modifiers will help you to avoid the problems described below.

Misplaced modifiers

Modifiers need to be put as close as possible to the words they refer to. If there is some distance between a modifier and the word it modifies, a reader may misinterpret the sentence.

✗ Students can access information about eating nutritiously <u>with this app</u>. [*Are they eating with the app?*]

✓ <u>With this app</u>, students can access information about eating nutritiously.

Be particularly attentive to words like *hardly, nearly, even, only,* and *almost,* which can modify many of the words in a sentence. Put them directly before the words they are meant to modify—just so that there is no room for misinterpretation:

✓ <u>Only</u> this study surveys the transformations.

✓ This study <u>only</u> surveys the transformations.

✓ This study surveys <u>only</u> the transformations.

✓ This study surveys the <u>only</u> transformations.

✓ This study surveys the transformations <u>only</u>.

Squinting modifiers

Remember that clarity largely depends on word order: to avoid confusion, make connections between the different parts of a sentence clear. Keep modifiers as close as possible to the words they modify. A *squinting modifier* is one that, because of its position, seems to be working in two directions at the same time:

✗ Resistors that malfunction <u>often</u> need to be replaced.

Does *often* refer to the rate of malfunction or the rate of replacement? Changing the order of the sentence or rephrasing it will make the meaning clearer:

✓ Resistors need replacing if they malfunction <u>often</u>.

✓ <u>Often</u>, resistors that malfunction need to be replaced.

Other ambiguous modifiers can be corrected in the same way:

✗ Dr. Hutt gave a lecture on nanorobotics, <u>which</u> has several extended applications.

✓ Dr. Hutt's <u>lecture</u> on nanorobotics <u>has</u> several extended applications.

✓ Dr. Hutt lectured on <u>nanorobotics</u>, <u>which</u> has several extended applications.

Split infinitives

Many readers object to the positioning of modifiers between the two parts of an infinitive (*to* and the base verb). Indeed, it's often more elegant to move the modifier to another position, either earlier or later, in the phrase. This revision is recommended whenever it is easy to make:

orig. To <u>graphically</u> represent time requires four coordinates.

rev. To represent time <u>graphically</u> requires four coordinates.

Dangling modifiers

Modifiers that have no grammatical connection with anything else in the sentence are said to be *dangling*. Sentences that begin with modifiers need special care. If it isn't clear who or what is doing the action expressed by the modifier, the dangling construction needs to be repaired:

✗ <u>Developing</u> a third model, <u>success</u> was eventually achieved.

It is unclear who is doing the developing. Here are two more examples:

✗ Before <u>setting out</u> to model the system, <u>it</u> is important to define the intended benefits.

✗ To <u>understand</u> the concept, <u>knowledge</u> of market risk is essential.

It is unclear who is doing the setting out or the understanding. Clarify meaning by putting the subject for the modifier close to it:

✓ <u>Developing</u> a third model, <u>the designers</u> finally achieved success.

✓ Before <u>setting out</u> to model the system, <u>the project team</u> must define the intended benefits.

✓ To <u>understand</u> the concept, <u>students</u> need a knowledge of market risk.

One type of dangling modifier occurs when the subject is hidden by a passive verb:

✗ To <u>maximize</u> the validity, a target user <u>group</u> has to <u>be designated</u>.

✓ To <u>maximize</u> the validity, <u>testers must designate</u> a target user group.

In another situation, changing an active verb in the modifier to a passive verb will solve the problem:

✗ After <u>modifying</u> the filter, it performed extremely well.

✓ After the filter <u>was modified</u>, it performed extremely well.

In conclusion, it makes sense to watch for problem modifiers when you are editing, for they occur far too frequently in scientific and other kinds of writing.

Problems With Pairs And Parallels

Comparisons

Make sure that your comparisons are complete. The second element in a comparison should be equivalent to the first, whether the equivalence is stated or merely implied:

✗ Today's students understand <u>calculus</u> better than <u>their parents</u>.

This sentence suggests that the two things being compared are *calculus* and *parents*. Adding a second verb (*do*) that matches the first one (*understand*) shows that the two things being compared are parents' understanding and students' understanding:

✓ Today's students <u>understand</u> calculus better than their parents <u>do</u>.

A similar problem arises in the following comparisons:

✗ A design engineer's <u>responsibilities</u> are similar to an <u>architect</u>.

✗ <u>The area</u> of the landfill site is twice as large as the proposed <u>development</u>.

In both cases, the writer is neglecting one of the primary principles of comparison—that the left side and right side must be equal and comparable:

✓ A design <u>engineer's</u> responsibilities are similar to an <u>architect's</u>.

✓ <u>The area</u> of the landfill site is twice as large as <u>that of</u> the proposed development.

Parallel phrasing

A series of items in a sentence should be phrased in parallel wording. Make sure that all the parts of a parallel construction (A, B, and C) are in fact equal:

✗ He liked <u>the pay,</u> <u>being able</u> to vary his hours, and also <u>appreciated</u> the many benefits.

✓ He liked <u>the pay,</u> <u>the flexible hours,</u> and <u>the many benefits.</u>

Once you have decided to use the same pattern in the first two elements, the third must have it as well. For clarity as well as elegance, keep similar ideas in similar form:

✗ The products are <u>extremely strong,</u> <u>dimensionally stable,</u> and <u>they do not contain formaldehyde.</u>

✓ The products are <u>extremely strong,</u> <u>dimensionally stable,</u> and <u>free of formaldehyde.</u>

The rule applies to lists as well, where bullets, with their implied *and*, call for parallel phrasing:

✗ Communications system use coding for three major reasons:
  - <u>reducing</u> the volume of information
  - <u>protection</u> against intruders
  - <u>to personalize</u> it

✓ Communication systems use coding for three major reasons:
- to <u>reduce</u> the volume of information
- to <u>protect</u> it against intruders
- to <u>personalize</u> it

Correlatives (coordinate constructions)

Constructions such as *both . . . and, not only . . . but also,* and *neither . . . nor* demand special care. The coordinating term must not come too early or else one of the parts that come after will not connect with the common element. For the implied comparison to work, the two parts that come after the coordinating terms must be grammatically equivalent:

✗ Mechanical systems should be not only <u>simple and reliable</u> but <u>require little maintenance</u>.

✓ Mechanical systems should not only <u>be</u> simple and reliable but <u>require</u> little maintenance.

Problems For English Language Learners

If English is not your first language, you have likely spent more time in English grammar and writing classes than the average student. The following explanations point out some of the most persistent trouble spots to watch for as you edit your work. When in doubt about vocabulary or idioms, consult the latest editions of the *Oxford Advanced Learner's Dictionary* and the *Oxford Collocations Dictionary*.

Noun rules
Use the determiner that's right for the context
Most nouns can be introduced by a word called a *determiner*. (The articles *a, an,* and *the* are the most common determiners, but others, called *quantifiers,* have similar functions.) Because nouns can be singular or plural, countable or uncountable, general or specific, it is usually the determiner that tells what kind of noun will follow. Choosing the appropriate determiner for the context will prevent you from sending mixed messages to your reader about the kind of noun you are using.

English nouns can be categorized in terms of quantity either by number or by amount. Singular countable nouns (such as *lake* or *contaminant*) are easily confused with uncountable nouns (such as *water* or *pollution*) because they both take singular verbs and because neither of them has an added *-s.* Plural countable nouns (*lakes, contaminants*) are easier to spot, but many nouns play dual roles, acting as countable or uncountable according to the context. It's the choice of determiners that makes the difference.

1. Use the determiners *a, an, one, another, each, every, either,* and *neither* only with singular countable nouns. If you remember the rule that every singular countable noun *must* have a determiner of some sort or other, you will avoid making errors like the one below:

   ✗ ITS encourages commuters to take bus rather than drive car.

   ✓ ITS encourages commuters to take <u>a</u> bus rather than drive <u>a</u> car.

2. Animals and people are always countable. It is a mistake not to use a determiner when they are used in the singular.

   ✗ The computer is often compared to the brain of human being.

   ✓ The computer is often compared to the brain of <u>a</u> human being.

3. Do not use singular countable determiners with uncountable nouns:

   ✗ <u>Every</u> slang or jargon must be avoided in formal writing.

   ✓ <u>All</u> slang or jargon must be avoided in formal writing.

4. Uncountable nouns don't need a determiner when they are used in a general sense. If you use a singular word without a determiner, you automatically tell the reader that it's uncountable. Be sure that's what you intend.

   ✓ Many people are afraid of <u>change</u>.

   ✗ There's been <u>change</u> in plans.

   ✓ There's been <u>a change</u> in plans.

5. The following determiners can all be used before plural (countable) nouns: *other, all, some, more, most, a lot of*:

other courses	all students
some books	more requirements
most colleges and universities	a lot of exams

   But these determiners are also used to introduce uncountable nouns:

other coursework	all software
some research	more evidence
most equipment	a lot of studying

Be careful not to confuse your singulars and plurals:

✗ Some lab need more spaces for machinery.

✓ Some labs need more space for machinery.

6. Watch out especially for nouns that can be either countable (one ____)
or uncountable (a lot of ____) according to the context. For example,
the word *experience* as a countable noun refers to a single eventful
occurrence: "Meeting the astronaut was quite *an experience*." When
referring to someone's background, however, use uncountable determiners: "He doesn't have *much experience*." Always use determiners
that clarify your meaning:

✗ Dr. Frank had a trouble solving the equation.

✓ Dr. Frank had some trouble solving the equation.

Words like *attention, difficulty, damage, effort, exercise, interest, life,
power, promise, proof, respect,* and *time* are only a few of those that
create similar difficulties for English language learners.

7. All numbers higher than *one* are used only with countable plurals
(*two eyes, three wheels, four legs*, and so on), as are the following quantifiers: *these, those, many, several, few, fewer, both*, and *a couple of*. Be
sure that the noun following them is indeed a plural. Sometimes, you
will have to use a countable word or phrase like *a piece of* together
with an uncountable noun:

✗ Imperial measures are still used in many equipment.

✓ Imperial measures are still used in many pieces of equipment.

8. The following quantifiers introduce uncountable nouns only: *much,
little, less,* and *an amount of*. You will make a mistake if you use one of
them in front of a plural:

✗ Channel coding produces less errors than expected.

✓ Channel coding produces fewer errors than expected.

9. The definite article *the*, the possessive pronouns (*my, your, his, her, its,
our, their, whose*), the interrogatives *what* and *which*, and the determiners *no* and *any* can be used before any noun at all. But when you use *the*,

you must already have a specific context for your noun. It is a mistake to use *the* when you mean *any*:

✗ The groundwater is an important source of water for agriculture.

✓ Groundwater is an important source of water for agriculture.

10. Before using a proper name, be sure you know whether it takes the determiner *the*. Nouns that are capitalized won't normally be introduced by *the* unless they also contain a phrase with *of*: it's *Vancouver*, but *the City of Vancouver*; *Waterloo College*, but *the University of Waterloo*; *Scotiabank*, but *the Bank of Montreal*. When the proper noun is included as a modifier, however (as in *the English language* or *the Calgary Flames*), you must use *the* for the sake of specificity:

✗ They went to a conference in United States.

✓ They went to a conference in the United States.

Verb rules

Watch agreement with singulars and plurals

Countable nouns can be singular or plural, where uncountable nouns have only a singular form. In addition to regular rules for subject–verb agreement, you must pay careful attention to the following situations.

1. With present-tense verbs and a *singular* subject, be sure to add *-s* to the verb:

   ✗ An RGP lens offer high oxygen transmissibility.

   ✓ An RGP lens offers high oxygen transmissibility.

   The verb must be singular as well when one of the following singular pronouns is the subject of the sentence: *one, each, either, neither, another, much, little, less*.

   ✗ Of the two sets of figures, neither are easy to manipulate.

   ✓ Of the two sets of figures, neither is easy to manipulate.

2. When referring to countable items in a set, you may use the prepositional phrase *of the* before the item to measure quantities. Remember that the word following *of the* will always be plural: "one of the *books*,"

"both of the *tests*," "some of the *results*." Remember also that the verb agrees with the word before *of*: "*one* of the books *is*," "*both* of the tests *are*," "*some* of the results *were*." Such constructions are worth checking twice to avoid errors:

✗ <u>One</u> of the most complicated <u>application</u> of AI <u>are</u> computer games.

✓ <u>One</u> of the most complicated <u>applications</u> of AI <u>is</u> computer games.

With sets of uncountables, the word following *of the* is always singular and the verb is singular as well: "*much* of the *test is* invalid"; "*some* of the *equipment was* missing."

With pronouns that can be either uncountable or plural (*all, any, more, most, some, none*), subject–verb agreement depends on whether the pronoun's referent is singular or plural: "*all* of the *material is* ready"; "*all* of the *materials are* ready."

3. In classifications, expressions like *type of*, *sort of*, and *kind of* are quite restrictive in terms of what follows: agreement will depend on whether the main noun is countable or uncountable. With countable nouns, agreement rules require everything to be all singular or all plural: "What *sort* of *label is* required?" "What *kinds* of *labels are* needed?"

With uncountable nouns, the expression can be singular or plural, but the noun itself must always remain singular:

✗ The program could handle various types of <u>informations</u>.

✓ The program could handle various types of <u>information</u>.

4. Rules for subject–verb agreement do not apply to modal auxiliaries (*can, could, shall, should, will, would, may, might, must*), which do not have separate singular and plural forms. If you aren't sure whether a subject is singular or plural, you will be safe if you can use a modal:

✓ The criteria will need validating.

Be sure, however, that you use the root form of the verb following a modal auxiliary:

✗ Later, the principal investigators <u>could discussed</u> the project.

✓ Later, the principal investigators <u>could discuss</u> the project.

Verb forms
Use _be_ and _have_ with care
_Be_ is the verb most often used in English—both as a main verb and as an auxiliary. _Have_ is the next most common one. It makes sense, then, to watch how you use these two verbs in particular.

Use continuous verbs for action in progress
_Be_ followed by the present participle forms the continuous verb form to emphasize action in progress at a point in time in the past ("she _was reading_"), present ("she _is reading_"), or future ("she _will be reading_"). Use the continuous to add the meaning of "being in the process of" to your context: "I am [in the process of] studying for my finals." Verbs that express states of being and sense perceptions instead of action do not take the continuous form, however. Among these are the verbs _appreciate, believe, contain, hear, intend, know, mean, need, own, possess, see, understand,_ and _want_.

When the verbs _be_ and _have_ are used in the continuous form, they have specific meanings: _be_ means "behave" ("Joshua _is being_ unusually cooperative today"), and _have_ refers to temporary duration ("He _is having_ dinner"; "she _is having_ a party"). Always be careful to choose the appropriate form.

> ✗ We assume that everyone is understanding the academic integrity policy.

> ✓ We assume that everyone understands the academic integrity policy.

Use perfect forms to express completion
Have followed by the past participle produces the perfect form, which suggests completion of an activity prior to a point in time in the past, present, or future. Use the present perfect rather than the past tense to bring things up to date:

> ✗ We will be interviewing all next week because seventeen candidates submitted impressive CVs.

> ✓ We will be interviewing all next week because seventeen candidates have submitted impressive CVs.

Be careful with passive forms
_Be_ followed by the past participle produces the _passive voice_ (see pp. 169–170), which shows the subject of the sentence _receiving_ the action of the verb rather than _doing_ it. Be sure to keep such relationships distinct:

✗ An atom <u>composes</u> of a nucleus surrounded by orbiting electrons.

✓ An atom <u>is composed</u> of a nucleus surrounded by orbiting electrons.

Differentiate your participles

Because *be* can be used as an auxiliary with both the past and present participles, it is especially important to choose the appropriate participle form for the context. In other words, always be sure to distinguish between *doing* an action (verb + *ing*) and *receiving* an action (verb + *ed*):

✗ I am <u>interesting</u> in applying for a position with your company.

✓ I am <u>interested</u> in applying for a position with your company. (The position interests me.)

Use only gerunds in prepositional phrases

Prepositions like *in, of, on, for,* and so on may be followed by only one possible form of a verb: the verb ending in *-ing*, which is called a *gerund*. Be sure to use only this form in prepositional phrases:

✗ Designers enhance the SNR either <u>by reduce</u> the noise effects or <u>by increase</u> the signal level.

✓ Designers enhance the SNR either <u>by reducing</u> the noise effects or <u>by increasing</u> the signal level.

An apparent exception to this rule is the word *to*, which may be followed by a gerund or the base form of a verb, depending on the context. When *to* is being used as a preposition, it must be followed by a gerund; when it is being used as part of an infinitive, however, it is followed by the base form of the verb. This double role explains the difference between "She used to do calculus" (but she no longer does) and "She is used to doing calculus" (so ask her to help you). Be sure that you know which *to* is called for by the context:

✗ I am looking forward <u>to meet</u> you.

✓ I am looking forward <u>to meeting</u> you.

✓ I hope <u>to meet</u> you soon.

Include required verbs

Don't leave out the auxiliaries or the main verbs that convey the meaning in the sentence:

> ✗ He <u>working</u> for that oil company since February.

> ✓ He <u>has been</u> working for that oil company since February.

Idiom rules

Idiom is the term used for a construction we can't explain except to say that it "sounds right." Idioms aren't predictable, and they aren't logical. The following are typical situations where you want to be careful that things sound the way they should.

Pay attention to word endings

As you add to your vocabulary, it's worth noting the suffixes that mark the various parts of speech (*-ment, -ness, -er, -ence* for nouns; *-ify, -ate, -en, -ize* for verbs; *-al, -ful, -ent, -like, -less* for adjectives; *-ly* for adverbs). Be sure to use the appropriate word for the context:

> ✗ Please send the following files at your <u>convenient</u>.

> ✓ Please send the following files at your <u>convenience</u>.

Be sure the suffix you have used is legitimate:

> ✗ This insulation is valued for its <u>strongness</u> and <u>durableness</u>.

> ✓ This insulation is valued for its <u>strength</u> and <u>durability</u>.

Include expected prepositions

1. Some prepositions (words such as *of, for, in, on,* and so on) express conventional relationships. The following sentences represent typical examples:

   > ✓ I am a student <u>in</u> the Faculty of Engineering and Applied Science <u>at</u> Queen's University <u>in</u> Kingston, Ontario.

   > ✓ I'll meet you <u>at</u> noon <u>on</u> Monday <u>at</u> the lab <u>on</u> King Street.

   Although *at* specifies time or place here and *on* generalizes, the relationships cannot be applied to other situations haphazardly:

   > ✗ This work will be presented <u>in</u> the graduate research conference <u>at</u> August 2–4, 2016.

    ✓ This work will be presented <u>at</u> the graduate research conference <u>on</u> August 2–4, 2016.

2. Verb or adjective constructions completed with prepositions don't usually allow for any choice. From *consist of* to *insist on* and *convenient to*, it is worth maintaining a list to refer to whenever you are writing.

Avoid unneeded prepositions

1. Do not include prepositions after verbs that do not require them:

    ✗ He lacked <u>of</u> motivation to find a job.

    ✓ He lacked motivation to find a job.

    Lack is used as a noun followed by *of* in the expression *have a lack of*. But because it uses four words instead of one, you should prefer the single verb *lack* for the sake of sentence economy.

2. With time expressions introduced by *last, next, this,* or *every* (*last night, next Monday, this month,* or *every week*), it is redundant to use a preposition:

    ✗ The midterm is <u>on</u> next Monday.

    ✓ The midterm is <u>next Monday</u>.

3. A few key idioms are completed by either gerund phrases or prepositional phrases:

    to have trouble/difficulty/a problem <u>doing</u> something
      or
    to have trouble/difficulty/a problem <u>with</u> a topic (*or* in a subject)

    to spend or waste time/money <u>doing</u> something
      or
    to spend or waste time/money <u>on</u> it

    to keep busy <u>doing</u> something
      or
    to keep busy <u>at</u> it.

    It's redundant to use the gerund and the preposition together:

    ✗ The professor spent three hours <u>on explaining</u> the project.

    ✓ The professor spent three hours <u>explaining</u> the project.

Use the idiomatic verb form called for by the main verb

1. Some verbs must be followed by an infinitive; others, by a gerund. Some (*begin, cease, continue, dread, forget, hate, like, love, prefer, remember, start, stop, try*) are followed by either form, sometimes with a change in meaning. Be sure you use the right idiom:

   ✗ Chemical reactions result when atoms <u>try sharing</u> electrons with other atoms.

   ✓ Chemical reactions result when atoms <u>try</u> to share electrons with other atoms.

2. There are eight verbs (*have, make, let, help, see, watch, hear,* and *feel*) that are followed by an object and the root form: "They *made* you *learn* this." Memorize these idioms, and avoid mistakes with them in your writing:

   ✗ Because of her high average, her faculty <u>let</u> her <u>to take</u> six courses in the winter term.

   ✓ Because of her high average, her faculty <u>let</u> her <u>take</u> six courses in the winter term.

3. Nine verbs (*ask, demand, insist, prefer, recommend, request, require, suggest, urge*) may be followed by a *that*-clause containing a verb in the subjunctive (the uninflected root form of the verb): "She *suggested that* he *be* consulted." Only four of these verbs (*ask, prefer, require, urge*) allow the clause to be rephrased with an infinitive phrase: "They urged her *to accept* the job." Do not use infinitive phrases with the other five verbs:

   ✗ She suggested <u>him to take</u> some time off before finding a job.

   ✓ She suggested <u>that he take</u> some time off before finding a job.

Watch word order

1. Word order in declarative sentences follows a basic pattern of subject followed by verb. The order is inverted in questions, whether they are information questions beginning with an interrogative (*What* is the density? *How long* does the reaction take?) or questions where the expected answer is "yes" or "no" (Is the frequency modulated? Did the program work?). When these questions are reported in sentences,

however, the original word order is preserved. Be careful, then, to respect the difference between *direct* and *indirect* questions:

✗ <u>Why</u> the e number <u>is</u> so important to scientists? (*direct*)

✓ <u>Why is</u> the e number so imporant to scientists? (*direct*)

✓ The student asked <u>why the e number is</u> so important to scientists. (*indirect*)

2. The subject–verb word order is also reversed when a sentence begins with a restrictive modifier introduced by *only* or a negative like *not*, *never*, or *seldom*. Regular word order applies if such modifiers appear later in the sentence:

✓ <u>Only once has</u> the computer lost a match.

✓ <u>Rarely does</u> the computer <u>lose</u> a match.

✓ The computer <u>rarely loses</u> a match.

The same principles apply to sentences introduced by *not only* and *neither* and/or *nor*. It is a mistake not to invert the subject and verb after these correlative conjunctions if you use them to join complete sentences:

✗ Not only <u>CNC performs</u> operations that people used to do, but it does work that was previously impossible to do.

✓ Not only <u>does</u> CNC <u>perform</u> operations that people used to do, but it does work that was previously impossible to do.

Avoid ungrammatical repetition

1. When you combine two sentences, use either coordination or subordination. It is redundant to use both:

✗ <u>Although</u> there are some limitations, <u>but</u> this method does identify clear trends.

✓ There are some limitations, <u>but</u> this method does identify clear trends. (*coordination*)

✓ <u>Although</u> there are some limitations, this method does identify clear trends. (*subordination*)

2. Avoid repetition in relative clauses. Once you've used a relative pronoun or adverb, be sure to delete the original referent:

✗ Seismologists are needed in countries <u>where</u> there have been major earthquakes <u>there</u>.

✓ Seismologists are needed in countries <u>where</u> there have been major earthquakes.

Distinguish between "it is" and "there is"

Despite sharing the main verb *be*, *it is* and *there is* (or *there are*) constructions are idiomatically quite different. They cannot substitute for each other: *It is time to work* is different from *There is time to work*. One test of whether it is appropriate to use *there is* is to try substituting *exists* for *is* (or *exist* for *are* if you are trying to use *there are*). If you can't make the substitution, try another idiom:

✗ <u>It is</u> a number of applications for packet filtering.

✓ <u>There are</u> a number of applications for packet filtering.

✓ <u>There exist</u> a number of applications for packet filtering. (*test*)

Be careful to include *it* or *there* in front of *is*. Even though neither word conveys much meaning, each one is essential idiomatically:

✗ <u>Is</u> theoretically possible to control traffic flow even in a large city.

✓ <u>It is</u> theoretically possible to control traffic flow even in a large city.

Avoid errors with comparisons

1. One- and two-syllable adjectives and some adverbs express the comparative by adding -*er* and a phrase beginning with *than*: *harder than x, easier than y, sooner than z*. Longer words and nouns show the comparison with *more* instead: *more difficult than x, more research than y*. It is redundant to use both methods of comparison together:

✗ Coding makes communication <u>more</u> faster and cheaper.

✓ Coding makes communication <u>much</u> faster and cheaper.

If you can use *more* in a comparison, then you can express the opposite with *less*, as long as you are comparing modifiers. (If you are using nouns, use *fewer* if the noun is plural and countable [see

p. 203].) If your comparison uses -er rather than *more*, you can express the opposite with the phrase *not as . . . as* instead.

✗ Plastics are <u>less hard</u> than ceramics but have higher impact resistance.

✓ Plastics are <u>not as hard as</u> ceramics but have higher impact resistance.

2. Don't forget to complete phrases including *so*, *such*, *too*, and *enough* for expressions of degree. Follow *so* or *such* with a *that*-clause: *The paper had so many errors that it was rejected*. With *too*, use an infinitive phrase: *It was too complicated to correct*. With *enough*, use either construction. In writing, it is an error not to complete such expressions. Revise to avoid the problem:

✗ Nature presents <u>so many</u> examples of symmetry.

✓ Nature presents <u>many</u> examples of symmetry.

3. Use standard phrases to establish comparisons: *different from* (not *than*), *similar to*, and *the same as*. Be sure that you always include *the* when you use the word *same*:

✗ At equilibrium, forward and reverse reactions occur at <u>same</u> rates.

✓ At equilibrium, forward and reverse reactions occur at <u>the same</u> rates.

Readers expect writing to be free of distracting or disruptive mistakes. Errors in grammar and usage, whether unconscious or careless, give the reader a negative impression. As you develop skills as a writer, you'll want to improve your editing and proofreading skills. Being aware of the most likely mistakes and learning to avoid them are worthwhile practices.

Chapter Checklist

☐ Keep a list of errors that have been pointed out to you, and look for these first when you proofread.

☐ Be wary of corrections suggested by your word processor's grammar checker if you don't understand the error.

☐ Consult your school's writing centre or its website for explanations of errors you have trouble understanding. Consider taking a course or attending sponsored workshops on campus.

☐ Book mark reliable websites like Purdue's Online Writing Lab or the Government of Canada's language portal for a quick link to resources.

Punctuation

Introduction

Punctuation poses enough problems that it deserves a chapter of its own. If your punctuation is flawed, your readers may be confused and forced to backtrack; worse, they may not be convinced that you are in control as a writer. Punctuation marks are the traffic signals of writing. Use them with precision to keep readers moving smoothly through your work.

Items in this chapter are arranged alphabetically: apostrophe, brackets, colon, comma, dash, ellipsis, exclamation mark, hyphen, italics, parentheses, period, quotation marks, semicolon, and slash.

Apostrophe [']

1. **Use an apostrophe to indicate the possessive form for nouns and some indefinite pronouns.**

   - Add an apostrophe followed by "s" to all singular and plural nouns not already ending in "s":

     Schrödinger's cat; women's studies.

   - Add an apostrophe followed by "s" to indefinite pronouns:

     no one's fault; anybody's guess

- Add an apostrophe followed by "s" to singular nouns and short names ending in "s" when you pronounce the new ending "iz":

  his boss's proposal; Ross's idea; Willis's autobiography

  Note, however, that you omit the second "s" when the word has more than than two syllables:

  Socrates' disciples; Copernicus' discovery

- Add only an apostrophe to plural nouns ending in "s", remembering to avoid confusing singulars and plurals:

  the Board of Directors' decision; several students' projects (but one student's project)

2. **Use an apostrophe before or after "s" (according to the rules above) to show time or distance measurements:**

   a month's notice; two weeks' time; a stone's throw

3. **Use an apostrophe to show contractions of words in informal writing:**

   we'll see; you're welcome; the '90s

   Caution! Don't confuse *it's* (the contraction of *it is* or *it has*) with *its* (the possessive of *it*), which has no apostrophe.

4. **Do not use an apostrophe to signal plurals of acronyms and numbers.**

   ✗ URL's were standardized in the 1990's.

   ✓ URLs were standardized in the 1990s.

Brackets [ ]

Brackets are square enclosures, not to be confused with parentheses (which are round).

1. **Use brackets to set off an editorial remark within a quotation.** They show that the words enclosed are not those of the person quoted:

> Before her marriage soured, Mileva Einstein Maric liked to joke that she and Albert were "one stone [ein Stein]."

When a direct quotation is unavoidable, use brackets to enclose *sic* after an error, such as a misspelling, to show that the mistake appeared in the original. (Most of the time you paraphrase to avoid reporting errors):

> Banting's discovery of insulin can be traced to brief jottings in his lab book: "Diabetus [*sic*]. Ligate pancreatic ducts of dog."

2. **Use brackets to indicate references in scientific writing.** Include the number of the citation in brackets when you make the reference in the text of your work:

> Dr. Knuth named the approach "literate programming" [3].

Then use the same number in brackets in the *References* section at the end to provide full documentation of the source. See p. 100.

3. **Use brackets within parentheses to indicate an additional parenthetical insertion** (This is the inverse of mathematical "fences"):

> The National Research Council (which delivers Canada's Business Innovation Access Program [BIAP]) encourages applications for short-term and long-term projects.

Colon [:]

A colon indicates that something will follow: an elaboration, a list, a quotation.

1. **Use a colon before a formal statement or series:**

> ✔ In physics, matter exists in four states: solid, liquid, gas, or plasma.

It is considered informal to use a colon if the words preceding it do not form a complete sentence:

✗ In physics, matter exists as: solid, liquid, gas, or plasma.

✓ In physics, matter exists as solid, liquid, gas, or plasma.

On the other hand, it is conventional to use a colon to introduce a vertical list, even when the introductory part is not a complete sentence:

✓ In physics, matter exists as:
  – solid
  – liquid
  – gas
  – plasma

Even so, professional writers usually recast the introductory phrase as a complete sentence.

2. **Use a colon for formality before a direct quotation, especially when introduced by a complete sentence:**

    If you seek success, remember the words of Thomas Edison: "Genius is 1% inspiration and 99% perspiration."

3. **Use a colon between numbers expressing time and ratios:**

    4:30 p.m.
    The ratio of calcium to potassium should be 7:1.

Comma [,]

Commas are the trickiest of all punctuation marks; even experts differ on when to use them. Most agree, however, that too many commas are as bad as too few, since they can make writing choppy and awkward to read. Certainly writers now use fewer commas. When in doubt, let clarity be your guide. The following are the most widely accepted conventions.

1. **Use a comma to separate two independent clauses joined by a co-ordinating conjunction (*and, but, for, or, nor, yet, so*).** By signalling that there are two clauses, the comma will prevent the reader

from confusing the beginning of the second clause with the end of the first:

✗ Pine is softer than oak and cedar is harder than oak.

✓ Pine is softer than oak, and cedar is harder than oak.

When the second clause has the same subject as the first, you have the option of omitting both the second subject *and* the comma:

✓ Rebuilding costs more at the outset, but it helps avoid expensive maintenance later.

✓ Rebuilding costs more at the outset but helps avoid expensive maintenance later.

If you mistakenly punctuate two sentences as if they were one, the result will be a run-on sentence (see pp. 183–184). If you use a comma but forget the coordinating conjunction, the result will be a comma splice:

✗ The Weil reaction of the capacitor is small, its impact is negligible.

✓ The Weil reaction of the capacitor is small, <u>so</u> its impact is negligible.

Remember that words such as *however, therefore,* and *thus* are transitional adverbs, not conjunctions. If you use one of them to join two independent clauses, the result will again be a comma splice:

✗ Enantiometers have the same molecular structures and physical properties, <u>however</u> they cannot be superimposed.

✓ Enantiometers have the same molecular structures and physical properties; <u>however</u>, they cannot be superimposed.

Transitional adverbs are often confused with conjunctions. You can distinguish between the two if you remember that a transitional adverb's position in a sentence can be changed:

✓ Enantiometers have the same molecular structures and physical properties; they cannot, <u>however</u>, be superimposed.

The position of a conjunction, on the other hand, is invariable; it must be placed between the two clauses:

✓ Enantiometers have the same molecular structures and physical properties, <u>but</u> they cannot be superimposed.

A good rule of thumb, then, is to use a comma when the linking word can't be moved.

2. **Use a comma between items in a series.** Place a comma and a co-ordinating conjunction before the last item:

✓ Coding is a procedure that makes communication faster, cheaper, more reliable, and more private.

✓ Software reengineering redesigns a system to improve its quality, understandability, and maintainability.

The comma before the conjunction is optional for single items in a series. Decide whether you'll include it or leave it out consistently:

✓ Fractals are found in vegetables, leaves(,) or snowflakes.

For phrases in a series, however, use the final comma to help prevent confusion:

✗ Linseed is a superior material because it is made from natural products (linseed oil, flax, jute and wood fibres).

Here a comma prevents the reader from thinking that *linseed oil* is made from *jute fibres* as well as *wood fibres*:

✓ Linseed is a superior material because it is made from natural products (linseed oil, flax, jute, and wood fibres).

This is the reason many writers always add a comma before the conjunction as soon as there are three or more items in a series. One rule is certain—remember not to put a comma at the end of the series if the series begins the sentence:

✗ Heat distribution, electric charge diffusion, wave propagation, and even chemical reactions, have mostly a natural exponential form.

✓ Heat distribution, electric charge diffusion, wave propagation, and even chemical reactions have mostly a natural exponential form.

3. **Use a comma to separate adjectives preceding a noun when they modify the same element and when you can substitute the co-ordinator *and* for the comma while retaining the same meaning.**

✗ It is a pretty clever design. [The design is quite clever.]

✔ It is a pretty, clever design. [The design is both clever and pretty.]

However, when the adjectives do not modify the same element, you should not use a comma:

✗ It is a familiar, domestic task.

Here *domestic* modifies *task*, but *familiar* modifies the whole phrase *domestic task*. A good way of checking whether you need a comma is to see if you can reverse the order of the adjectives. If you can (*pretty, clever device* or *clever, pretty device*), use a comma; if you can't, omit the comma:

✗ It is a domestic familiar task.

✔ It is a familiar domestic task.

4. **Use commas to set off an interruption (i.e., "parenthetical element"):**

✔ The results, as expected, were inconclusive.

✔ It is important, however, that results be consistent.

Remember to put commas on both sides of the interruption:

✗ It is important however, that results be consistent.

✗ The build-up of these fumes, along with mold, bacteria, and dust can lead to "sick-building syndrome."

✔ The build-up of these fumes, along with mold, bacteria, and dust, can lead to "sick building syndrome."

5. **Use commas, like parentheses, to set off words or phrases that provide additional but non-essential information:**

✔ Our TA, Chandra Elliot, gives clear explanations.

✔ The video clip, a documentary on recycling, was popular among students.

In these examples, *Chandra Elliot* and *a documentary on recycling* give additional information about the nouns they refer to (*TA* and *video clip*), but the sentences would make sense without them. Here's another example:

✔ Polymers involve macromolecules, which have high molecular weights.

The phrase *which have high molecular weights* is a *non-restrictive modifier* because it doesn't limit the meaning of the word it modifies (*macromolecules*). Without that modifying clause, the sentence would still make sense. Since the information the clause provides is not necessary to the meaning of the sentence, you use a comma to set it off.

In contrast, a *restrictive modifier* is one that provides essential information. It must not be set apart from the element it modifies, and commas should not be used:

✔ Many people who are legally blind can read using a closed circuit TV.

Without the clause *who are legally blind*, the reader would not know which specific people are being referred to. More importantly, the addition of commas suggests that many people are legally blind—an error not to make.

To avoid confusion, be sure to distinguish carefully between essential and additional information. The difference can be important:

✗ Students, who are unwilling to work, should not receive grants.

✔ Students who are unwilling to work should not receive grants.

The first example makes an unacceptable generalization about all students, which you understand if you imagine parentheses in place of the commas.

The issue is never simple, as you can see above. In the example that follows, adding or omitting a comma changes the meaning just as significantly:

✔ There is really only one problem which calls for immediate action.

To avoid misinterpretation in such cases, it makes sense to follow the practice of using *which*, following the comma, only with non-restrictive (i.e., non-essential) modifiers; use *that* (without the comma) to introduce restrictive modifiers:

✔ There is really only one problem that calls for immediate action. (*restrictive*)

✔ There is really only one problem, which calls for immediate action. (*non-restrictive*)

6. **Use a comma after an introductory phrase, especially when omitting it would cause confusion:**

   ✗ After extensive planning tests were performed with real data.

   ✓ After extensive planning, tests were performed with real data.

   ✗ If the samples are random calculations need adjusting.

   ✓ If the samples are random, calculations need adjusting.

7. **Use a comma to separate elements in titles, dates, and addresses:**

   David Gunn, President

   February 2, 2016 (Commas are generally omitted if the day comes first: 2 February 2016)

   117 Hudson Drive, Edmonton, Alberta

   They lived in Dartmouth, Nova Scotia.

8. **Use a comma before a quotation in a sentence:**

   Einstein said, "God is subtle, but he is not malicious."

   For more formality, if the quotation is preceded by a grammatically complete sentence, use a colon (see p. 217).

9. **Do not use a comma between a subject and its verb:**

   ✗ The metals and other ions in the contaminated groundwater, increase the water's conductivity.

   ✓ The metals and other ions in the contaminated groundwater increase the water's conductivity.

10. **Do not use a comma between a verb and its object:**

    ✗ The second section explains, what solutions are possible.

    ✓ The second section explains what solutions are possible.

11. **Do not use a comma before "and" or "or" when linking pairs of words, phrases, or dependent clauses:**

✗ In the simulation, two robots think independently, and cooperate to achieve a common goal.

✓ In the simulation, two robots think independently and cooperate to achieve a common goal.

12. **Do not use a comma in place of the colon to introduce a list:**

✗ Two AI tools were considered as possible components, natural language processors and neural networks.

✓ Two AI tools were considered as possible components: natural language processors and neural networks.

Dash [—]

A dash abruptly and dramatically draws attention to the words that follow. Never use dashes as casual substitutes for other punctuation. Overuse can detract from the calm, well-reasoned effect you want to create.

1. **Use a dash to stress a word or phrase:**

Pressure is a physics concept crucial to medicine—especially in lung physiology.

Their solution was well received—at first.

2. **Use a pair of dashes to set off an important interruption:**

Hawking postulates that if the universe has no boundary—no beginning and no end—then it is self-contained.

Note the effect of parentheses in the same context. Dashes emphasize; parentheses add an aside:

Hawking postulates that if the universe has no boundary (no beginning and no end) then it is self-contained.

Be careful to distinguish between hyphens and dashes. Two hyphens typed together with no spaces on their side represent a dash. This is the most space-efficient way of showing the feature, and word processors automatically convert this to a dash (called an *em dash* because of its length) as you continue typing. Another method is to use an *en dash*, one which is slightly longer than a hyphen, with single spaces left on either side.

Ellipsis [. . .]

1. **Use ellipsis points (three spaced dots) to show an omission from a quotation:**

    > She reported that "to many farming families in the West, the drought in the Thirties [. . .] resembled a biblical plague, even to the locusts."

    You can use brackets around the ellipsis to indicate that you, not the original author, have left out the words. If the omission comes before or after the words quoted, ellipsis points are not used:

    > She reported that the drought "resembled a biblical plague, even to the locusts."

    > She reported that the drought "resembled a biblical plague."

2. **Use ellipsis points to show that a series of numbers continues indefinitely:**

    $$y = 1, 3, 5, 7, 9, \ldots$$

    In mathematical copy, put commas or operational signs after each term and after the ellipsis points (without an intervening space) if followed by a final term:

    $$x_1, x_2, \ldots, x_n$$

Exclamation Mark [!]

An exclamation mark helps to show emotion or feeling, usually in informal writing such as e-mails. Use it only in those rare cases, in non-scientific writing, when you want to give a point emotional and dramatic emphasis:

> Prof. Furness predicted that the dollar would rise against the euro. Some forecast!

Hyphen [-]

1. **Use a hyphen if you are forced to divide a word at the end of a line.** Although it's generally best to start a new line if a word is too long, there are instances—for example, when you're formatting text in narrow columns—when hyphenation might be preferred. The hyphenation

feature in word processors has taken the guesswork out of dividing words at the end of the line, but if you must use manual hyphenation, here are a few guidelines:

- Divide between syllables.
- Never divide a one-syllable word.
- Never leave one letter by itself.
- Divide double consonants except when they come before a suffix, in which case divide before the suffix:

> col-lect
> fall-ing
> pass-able

When the second consonant has been added to form the suffix, keep it with the suffix:

> refer-ral
> begin-ning

- Do not, however, divide a hyphenated compound word except at the hyphen:

> ✗ co-op-erative

2. **Use a hyphen to separate the parts of certain compound words:**

- compound nouns:

> kilowatt-hour; brother-in-law

- compound verbs:

> test-drive; mass-produce

- compound modifiers:

> matrix-based consideration; first-aid kits

Do not, however, hyphenate a compound modifier that includes an adverb ending in -*ly*:

✗ a highly-developed prototype

✔ a highly developed prototype

Spell-checking features today will help you determine which compounds to hyphenate, but there is no clear consensus, even from one dictionary to another. As always, the solution for your writing style is for you to be consistent.

3. **Use a hyphen with certain prefixes (*all-*, *self-*, *ex-*, *e-*) and with prefixes preceding a proper name.** Practices do vary, so consult a dictionary when in doubt.

   all-inclusive; self-imposed; ex-president; e-commerce

Use *former* rather than *ex-* in formal situations:

   Dr. Ling is the former chair of this department.

4. **Use a hyphen to emphasize contrasting prefixes:**

   The technician assessed both pre- and post-test findings.

5. **Use a hyphen to separate written-out compound fractions and compound numbers from one to ninety-nine:**

   seven-tenths full; two-thirds of a cup; twenty-two participants

6. **Use a hyphen to join a cardinal number and the unit of measurement when they precede what they modify:**

   a five-step process; 16-point type; a six-legged robot

7. **Use an en dash rather than a hyphen (see p. 223) to separate parts of inclusive numbers or dates:**

   the years 1890–1914; pages 3–10

Italics [*Italics*]

1. **Use italics for the titles of books, journals, magazines, plays, films, and lengthy musical pieces, as well as for the names of ships, trains, and spacecraft:**

   The Discipline of Design is one of my favourite textbooks.

Note: for articles, chapter titles, and unpublished dissertations (as well as short poems or musical pieces), use quotation marks. If the title itself contains the title of another work, be sure to set it off in appropriate fashion:

- When both titles are those of major works, use quotation marks for the internal one:

    His latest documentary is *"Columbia"—What Went Wrong?*

- When the internal title is a major work but the main title is not, use italics:

    For more detail, see her recent article, "Special Effects in *Avatar*."

- When neither title is a major work, use single quotation marks for the internal title:

    The article is entitled "Assonance in 'The Hollow Men.'"

2. **Use italics (or quotation marks) to identify a word or phrase that is itself the subject of discussion:**

    The term *op amp* is one of many blended words used in electrical engineering.

3. **Use italics for foreign words or expressions that have not been naturalized in English, including Latin names for species and subspecies:**

    All owls belong to the sub-order *Striges*.

4. **Use italics for the text of equations as well as for defining theorems, rules, etc.:**

    $E = mc^2$

Parentheses [( )]

1. **Use parentheses to enclose an explanation, example, or qualification.** Parentheses show that the enclosed material is of incidental importance

to the main idea. They make an interruption that is more subtle than one marked off by dashes but more pronounced than one set off by commas:

✓ The frequency domain brings out the dynamic (non-random) characteristics of a signal.

✓ Only clusters with a population of greater than 3,000 (30% of the entire image) are processed.

Remember that punctuation should not precede parentheses but may follow them if required by the sense of the sentence:

✓ In cases that require special formatting (smoothed edges, shading, or shadowing), the characters must be modified.

If the parenthetical statement comes between two complete sentences, it should be punctuated as a sentence, with the period, question mark, or exclamation mark *inside* the parentheses:

✓ The process requires substantial execution time. (Turnaround can take as much as 24 hours.) Nevertheless, the savings outweigh the costs.

2. **Use parentheses to translate acronyms the first time you use them:**

✓ There are different types of FACTS (Flexible AC Transmission Systems) controllers.

✓ The concept explains how the central nervous system (CNS) operates.

3. **Use parentheses for in-text author or year references.** See Chapter 9 for details.

Period [.]

1. **Use a period at the end of a sentence.** A period indicates a full stop, not just a pause.

2. **Use a period with abbreviations.** Canada's adoption of the metric system contributed to a trend away from the use of periods in many

abbreviations. It is still common to use periods in abbreviated names and titles (Mr. M. J. Hunt, Rev. M. Collings-Moore etc.) and expressions of time (6:30 p.m.). Convention no longer requires them in academic degrees (MSc, PhD, etc.), and abbreviated names of states and provinces (BC, PEI, NY, DC). In addition, most acronyms for organizations do not use periods (CIDA, CBC, UNESCO, WTO), nor do general acronyms (RPM, GST, EMF, STP).

3. **Use a period at the end of an indirect question.** Do not use a question mark:

   ✘ Users were asked if they found the manual useful?

   ✔ Users were asked if they found the manual useful.

4. **Use a period for questions that are really polite statements:**

   ✔ Would you please send him the report by Friday.

   ✔ May I congratulate you on your promotion.

Quotation Marks [" " or ' ']

American and British methods of punctuating quotations differ. American practice generally favours double quotation marks, while British practice generally favours single quotation marks. Sometimes the decision is based on space constraints (as happens with newspaper headlines, which use only single quotation marks). In Canada, either style is accepted as long as you are consistent. The guidelines outlined below are based on the American conventions, which are more common in Canada than the British ones.

1. **Use quotation marks to signify direct discourse (the actual words taken from a speaker or a text):**

   In 1912, Einstein wrote the following to a friend: "Compared with this problem [formulating the general theory of relativity], the original theory is child's play."

2. **Use quotation marks to show that words themselves are the issue:**

   The tennis term "love" comes from the French word for "egg."

Alternatively, you may italicize the terms in question.

Only in informal writing are quotation marks occasionally used to show that the writer is aware of the difficulty with a slang word or inappropriate usage. You mark text this way only if you would use your fingers to make imaginary quotation marks when you are face to face with someone:

> Several of the "experts" did not seem to know anything about the topic.

In general, it's best to avoid sarcasm in any professional context. Remember the expectation of a professional attitude, however you may feel.

3. **Use quotation marks to enclose the titles of chapters, poems, short stories, songs, presentations, and articles in books or journals.** (In contrast, titles of books, films, paintings, and music are italicized.)

> The most interesting article in the June 2014 issue of *IEEE Potentials* is "What's behind your smartphone icons?"

4. **Use single quotation marks to enclose quotations within quotations:**

> He said, "Several of the 'experts' did not know anything about the topic."

Note that British style reverses this pattern so that double quotation marks enclose quotations within quotations.

Placement of punctuation with quotation marks

The following guidelines for punctuating quotations are based on American rather than British conventions:

- A comma or period always goes <u>inside</u> the quotation marks:

  > The method involves identifying the "core business," the "extended enterprise," and the "business ecosystem."

- A semicolon or colon always goes <u>outside</u> the quotation marks:

  > Cybernetics relates to the developing "sciences of complexity": AI, neural networks, and complex adoptive systems.

- A question mark, dash, or exclamation mark goes inside quotation marks if it is part of the quotation but outside if it is not:

  > One of the more important development questions is this: "How much investment money can be raised?"

  > Did Prof. Hagen actually describe the Casuistic Theory of Value as "passing the buck"?

- If a reference is given parenthetically (in round brackets) at the end of a quotation, the quotation marks precede the parentheses, and the sentence punctuation follows them:

  > Lipsey suggested that we should "abandon the Foreign Investment Review Agency" (Paisley 94).

Remember that questions, exclamation marks, and dashes are rare in scientific writing, but they have their place in informal communications.

Semicolon [;]

1. **Use a semicolon to join independent clauses (complete sentences) that are closely related:**

   > Some samples contained sediment; others were clear.

   A semicolon is especially useful when the second independent clause begins with a transitional adverb, such as *however, moreover, consequently, nevertheless, in addition,* or *therefore* (usually followed by a comma):

   > Daylight is the most efficient source of building lighting; however, its full potential has yet to be demonstrated.

   Some editors disagree, but it's usually acceptable to follow a semicolon with a coordinating conjunction if one or both clauses are complicated by other commas:

   > On July 14, 2015, staff participated in a teambuilding exercise; but the effects were, unfortunately, short-lived.

2. **Use a semicolon to mark the divisions in a complicated series when individual items themselves need commas.** Using a comma to mark

the subdivisions and a semicolon to mark the main divisions will help to prevent mix-ups:

✗ He invited Professor Ludvik, the vice-principal, Christine Li, and Dr. Hector Jimenez.

Is the vice-principal a separate person?

✓ He invited Professor Ludvik, the vice-principal; Christine Li; and Dr. Hector Jimenez.

A simpler alternative is to retain the commas and use parentheses around the secondary material:

✓ He invited Professor Ludvik (the vice-principal), Christine Li, and Dr. Hector Jimenez.

Slash [/]

1. **Use the slash to offer alternatives when either of a pair of words is to be selected. There is no spacing on either side of the slash:**

   Addresses should include name, street, city, province/state, country, and postal code.

   While *he/she* and *s/he* are gaining in popularity as gender-neutral pronoun pairs (see p. 243), be wary of using them in any formal writing. Unless space is an issue, write these out in full: *he or she*.
   Do not confuse the slash [/] with the backslash [\]. The backslash is generally only used in computing and should never be used to distinguish between alternatives.

2. **Use a slash to separate parts of a URL:**

   http://www.oupcanada.com/higher_education.html

The conventions of punctuation may not seem logical, but the rules are easy to follow once you know them. Take the time to become familiar with expectations, and then apply the rules consistently throughout your writing.

Chapter Checklist

☐ Proofread one sentence at a time to confirm that internal punctuation is used conventionally.

☐ Consult your school's writing centre or its website for exercises and explanations if you need them. Consider attending sponsored punctuation workshops on campus.

☐ Book mark reliable websites like Purdue's Online Writing Lab or the Government of Canada's language portal for a quick connection to resources.

Misused Words and Phrases

Introduction

This chapter offers an alphabetical overview of words and phrases that are often confused or misused. A periodic read-through will refresh your memory and help you avoid mistakes.

a, an. When you want to generalize about a single item, these are alternatives to the word *one*. Use **a** before a consonant sound and **an** before a vowel sound (*a, e, i, o, u*) or silent *h*:

> This product has a history of recalls.

> It will take an hour to repair.

accept, except. Accept is a verb meaning *to receive affirmatively*; **except**, when used as a verb, means *to exclude*:

> She accepted the scholarship.

> The course was designed for upper-year students, non-majors excepted.

acoustics. This word is singular as it refers to the science, but plural with respect to the acoustic properties of a building:

> Acoustics looks at how a building's acoustics are planned.

advice, advise. **Advice** is a noun, **advise** a verb:

> He was <u>advised</u> to ignore the <u>advice</u> of others.

affect, effect. **Affect** is a verb meaning *to influence*. **Effect** can be either a noun meaning *result* or a verb meaning *to bring something about*, usually with reference to change:

> The eye drops <u>affect</u> his vision.
>
> The <u>effect</u> of higher government spending is higher inflation.
>
> One job of the production engineer is to <u>effect</u> improvements.

all ready, already. To be **all ready** is simply to be ready for something; **already** means *beforehand* or *earlier*:

> The students were <u>all ready</u> for the test.
>
> Three days later, the test had <u>already</u> been marked.

all right. Write as two separate words: *all right*. (The single word *alright* is for informal writing only.) **All right** can mean *safe and sound, in good condition, okay, correct*, or *satisfactory*:

> Is everyone <u>all right</u>?
>
> The student's answers were <u>all right</u>.

(Does this second example mean that the answers were all correct or simply acceptable? Use a clear synonym instead.)

all together, altogether. **All together** means in a group; **altogether** is an adverb meaning *entirely*:

> He was <u>altogether</u> insistent on keeping the papers <u>all together</u>.

allusion, delusion, illusion. An **allusion** is an indirect reference; a **delusion** is a belief or perception that is distorted; an **illusion** is a false perception of reality:

> Her joke about relatives was an <u>allusion</u> to Einstein.

He had <u>delusions</u> that he could pass the exam without studying.

What looks like water on the road is often an optical <u>illusion</u>.

a lot. Write as two separate words: *a lot.*

alternate, alternative. Alternate means *every other* or *every second* thing in a series; **alternative** refers to a *choice* between options:

The two sections of the class attended labs on <u>alternate</u> weeks.

The students could do a research paper as an <u>alternative</u> to writing the exam.

among, between. The general rule is to use **among** for three or more persons or objects considered collectively and **between** for two (or more when used individually):

<u>Between</u> you and me, there's trouble <u>among</u> the committee members.

amount, number. Note the difference between what's countable and not countable when choosing between these two words. **Amount** indicates quantity when units are not discrete and not absolute (i.e., uncountable); **number** indicates quantity when units are discrete and absolute (i.e., countable):

A large <u>amount</u> of electricity was consumed.

A large <u>number</u> of students were registered.

See also **less, fewer.**

analysis. The plural is **analyses.**

ante-, anti-. Distinguish the spellings here. **Ante-** means *before*, as in *antecedent* or *antedate.* **Anti-** means *opposite* or *against*, as in *antidote* or *antimatter.*

anyone, any one. Write the two words to give numerical emphasis on **one**; otherwise, **anyone** (and its informal synonym anybody) is written as one word:

Any one of the proposals is reasonable.

Anyone can write a reasonable proposal.

anyways, anywheres. Non-standard informal English. Use **anyway** or **anywhere** instead.

as, because. As a synonym of *because*, **as** should be avoided because it may be confused with *while* or *when*:

✗ As he was working, he ate at his desk.

✓ Because he was working, he ate at his desk.

✗ She arrived as he was leaving.

✓ She arrived when he was leaving.

aspect, respect. Distinguish carefully between these two nouns when **aspect** means *angle* and **respect** means *point*:

This aspect of the problem needs expansion in two respects.

bad, badly. Bad is an adjective meaning *not good*:

The weather turned bad.

He felt bad about turning in his assignment late.

Badly is an adverb meaning *not well*. When used with the verbs **want** or **need**, it means *very much*:

The results were recorded badly.

The results badly needed corroboration.

basis, bases. **Basis** is singular; **bases** is plural.

beside, besides. **Beside** is a preposition meaning *next to*:

She worked beside her assistant.

Besides has two uses: as a preposition it means *in addition to*; as a transitional adverb it means *moreover*:

> Besides her assistant, there was no one in the lab.

> She left because she was tired. Besides, there was no more work to do.

between. See **among**.

bring, take. Use **bring** for action coming closer to the speaker (*here*) and **take** for action going away (*there*):

> Bring me your resumé the next time you come.

> Take your resumé when you go to your interview.

centrifugal, centripetal. A **centrifugal** force is directed away from the axis of rotation, where a **centripetal** force is directed toward the axis of rotation.

cite, sight, site. To **cite** something is to *quote* or *mention* it as an example or authority; **sight** relates to vision or to views; **site** refers to a specific *setting* or *location*, as in *website*.

> The tourists hopped on a bus to see the sights.

> Visitors must put on work boots and hard hats before touring the work sites.

classic, classical. Use **classic** to mean *memorable* or *standard of excellence*. **Classical** is normally the choice in scientific writing (as well as music):

> Gaussian distribution predicts the classic bell curve.

> Classical physics does not account for black holes.

climactic, climatic. Climactic describes a *climax*; **climatic** refers to *climate*.

complement, compliment. As verbs, **complement** means *to complete* or *enhance*, while **compliment** means *to praise*. Make this distinction clear, especially when you use adjective endings (**-ary**):

The two experiments produced complementary results.

The professor's comments on the report were complimentary.

New subscribers were promised complimentary apps.

compose, comprise. Both words mean *to constitute or make up*, but **compose** is preferred. **Comprise** is correctly used to mean *include, consist of*, or *be composed of*. Using **comprise** in the passive ("are comprised of")—as is tempting in the second example below—is frowned on in formal writing, especially because it uses three words where one will do:

Four students compose the group representing the faculty.

All systems comprise rules that govern them.

continual, continuous. **Continual** means *repeated over a period of time*; **continuous** means *constant* or *without interruption*:

The strikes caused continual delays in building the road.

Five days of continuous rain delayed the project further.

council, counsel. **Council** is a noun meaning *an advisory or deliberative assembly*. **Counsel** as a noun means *advice* or *lawyer*; as a verb it means *to give advice*.

The college council meets on Tuesdays.

A camp counsellor may need to counsel parents as well as children.

criterion, criteria. A **criterion** is *a standard for judging something*. **Criteria** is the plural of *criterion* and thus requires a plural verb:

These are the criteria for evaluating the new product.

data. The plural of **datum**. **Data** refers to the set of information, usually in numerical form, that is used for analysis as the basis for a study. Informally, **data** is often used as a singular noun, but in formal scientific contexts, treat it as a plural:

These data were gathered in random fashion. Therefore they are inconclusive.

deduce, deduct. To **deduce** something is *to work it out by reasoning*; to **deduct** means *to subtract or take away* from something. The noun form of both words is **deduction**.

defence, defense. Both spellings are correct: **defence** is standard in Britain and is somewhat more common in Canada; **defense** is standard in the United States.

defuse, diffuse. Although they sound similar, these words have quite different meanings. It is a bomb that you **defuse** and information that you **diffuse** (or send out).

delusion. See **allusion**.

dependent, dependant. **Dependent** is an adjective meaning *contingent on* or *subject to*; **dependant** is a noun.

> Suriya's graduation is dependent upon her passing algebra.
>
> Chedley is a dependant for income tax purposes.

device, devise. The word ending in **-ice** is the noun; the word ending in **-ise** is the verb.

> Once he had devised the new fastener, he patented the device.

different from, different than. Although **different than** is common in informal contexts, use **different from** in writing:

> The results were different from the predictions.

diminish, minimize. To **diminish** means *to make or become smaller*; to **minimize** means *to reduce* something to the smallest possible amount or size.

discreet, discrete. **Discreet** means *tactful* or *prudent*; **discrete** means *separate and distinct*, generally the meaning applicable in scientific writing:

> The interviewer was discreet in asking about previous jobs.
>
> Discrete samples were collected.

disinterested, uninterested. Disinterested implies impartiality or neutrality; **uninterested** implies a lack of interest:

> As a disinterested observer, she was able to judge the issue fairly.
>
> Uninterested in the proceedings, he yawned repeatedly.

dominant, dominate. Dominant is an adjective meaning *exerting control over* or *ranking above* something. **Dominate** is a verb meaning *to rule*:

> The dominant values are reported in Fig. 1.
>
> Safety should dominate all construction decisions.

due to. Although increasingly used to mean *because of*, **due** is an adjective and therefore needs to modify something:

> ✗ Due to his impatience, we lost the contract. [*Due is dangling.*]
>
> ✓ Because of his impatience, we lost the contract.
>
> ✓ The loss was due to his impatience.

economic, economical. Use **economic** in reference to *the economy* and **economical** in reference to *savings*.

e.g., i.e. E.g. means *for example*; **i.e.** means *that is*. It is considered incorrect to use them interchangeably.

eminent, imminent. Eminent means *prominent* or *distinguished*; **imminent** refers to time and means *soon*.

entomology, etymology. Entomology is the study of insects; **etymology** is the study of the derivation and history of words.

especially, specially. Especially means *particularly* where **specially** means *for a special purpose*:

> The device was specially designed for wheelchair users.
>
> Cost price is an especially important consideration in manufacturing.

etc. Avoid **etc.** (*et cetera*) in formal writing. End a list with two or three examples introduced by *such as*:

    ✗   The project leaders discussed risk, interest rates, <u>etc</u>.

    ✓   The project leaders discussed <u>such</u> variables <u>as</u> risk and interest rates.

everyday, every day. Both words mean *daily*, but **everyday** is an adjective (like *regular*), and **every day** is an adverb (like *regularly*):

    Sneakers are now appropriate for <u>everyday</u> wear.

    He wears sneakers <u>every day</u>.

exceptional, exceptionable. Exceptional means *unusual* or *outstanding*, whereas **exceptionable** means *open to objection* and is generally used in negative contexts:

    His accomplishments are <u>exceptional</u>.

    There is nothing <u>exceptionable</u> in his behaviour.

farther, further. Where **farther** generally refers to distance, **further** suggests extent:

    The aquifer lies <u>farther</u> south than expected.

    She explained the proposal <u>further</u>.

firstly. Use **first** or **first of all** to avoid annoying readers who consider **firstly** old-fashioned or pretentious.

flaunt, flout. Although they sound similar, these verbs have opposite meanings. **Flaunt** means *to show* something *off*, where **flout** means *to treat* something *with contempt or defiance*.

    The starlet <u>flouted</u> good taste by <u>flaunting</u> her diamonds.

focus. The plural of the noun may be either **focuses** (also spelled *focusses*) or **foci**.

foreword, forward. A **foreword** (or *preface*) refers to front matter in a book or dissertation; **forward** describes a direction:

> In his foreword, the author praised his team for continually moving the project forward.

good, well. Good is an adjective that modifies a noun; **well** is an adverb that modifies a verb. In medical contexts, however, **well** is an adjective meaning *healthy*.

> He is a good rugby player.
>
> The experiment went well.
>
> The patient reported feeling well.

hardy, hearty. To distinguish these sound-alike words, reserve the sense of *durable* for **hardy** and *substantial* or *energetic* for **hearty**:

> Corn is a hardy crop.
>
> Corn chowder makes a hearty meal.

he/she, his/her. In formal writing, do not resort to these abbreviations. If it is impossible to avoid generalizing in the singular, write out **he or she**, or the equivalent pronoun forms, in full.

> ✗ A co-op student should make sure his/her resumé is always up to date.
>
> ✓ A co-op student should make sure his or her resumé is always up to date.
>
> ✓ Co-op students should make sure their resumés are always up to date.

hopefully. In formal writing, do not use **hopefully** at the beginning of a sentence as a substitute for *it is to be hoped that.*

> ✗ Hopefully the results will be published.
>
> ✓ The investigators hope to publish the results.

i.e. This is not the same as **e.g.** See **e.g.**

illusion. See **allusion.**

incite, insight. **Incite** is a verb meaning *to stir up*; **insight** is a noun meaning (often sudden) *understanding.*

infer, imply. To **infer** means *to deduce or conclude by reasoning.* It is often confused with **imply**, which means *to suggest* or *insinuate.* Note the give-and-take in these words (we **infer** X from Y, where Y **implies** X to us):

> We infer from the data that there will be cost overruns.

> The data imply that there will be cost overruns.

inflammable, flammable, non-flammable. Despite its **in-** prefix, **inflammable** is not the opposite of **flammable**: both words describe things that are *easily set on fire.* The opposite of **flammable** is **non-flammable**. To prevent any possibility of confusion, most people have stopped using **inflammable** altogether.

inset, insert. Something that is **inset** is literally "set in," as a picture within a picture. Don't confuse this word with **insert**, a noun referring to something added separately:

> The histogram appears as an inset on the map of Cape Breton.

> A weekly regional paper comes as an insert in the *Saturday Record.*

intense, intensive. While both words refer to heavy concentrations, **intense** is a more subjective word, reflecting one's response to or impression of something; **intensive** is used as a more objective description:

> He found the course intense and had difficulty keeping up with the workload.

> It was an intensive course, designed to cover a lot of material in a short period of time.

irregardless. **Irregardless** may be found in dictionaries, but it is considered non-standard English. It is also redundant. Use **regardless** instead.

its, it's. Its is a possessive pronoun; **it's** is a contraction of *it is* or *it has*. It is an error to put an apostrophe anywhere in *its* to show possession:

✗ Every problem has its' solution.

✗ Every problem has it's solution.

✓ Every problem has its solution.

✓ It's time to leave.

lead, led. **Lead** (the metal) sounds like **led** (the past form of the verb **to lead**). Be careful to distinguish the spellings.

✗ The investigation lead to a public inquiry in 2013.

✓ The investigation led to a public inquiry in 2013.

✓ The investigation will lead to a public inquiry later this year.

less, fewer. **Less** is normally used with singular and uncountable units (as in "less information"). **Fewer** is the word to use with plurals (as in "fewer details"). However, **less** is also regularly used as a pronoun measuring statistics, distances, sums of money, or units of time, which are often thought of as amounts:

Abstracts should be kept to 150 words or less.

lie, lay. To **lie**, an intransitive verb, means *to assume a horizontal position*; to **lay**, a transitive verb, means *to put [something] down*. The changes of tense often cause confusion:

Present	Past	Past participle
lie (recline)	lay	lain
lay (put)	laid	laid

✗ The doctor told the patient to lay down.

✓ The doctor told the patient to lie down.

✓ The patient lay down as requested.

like, as. Like is a preposition, but it is often used casually as a conjunction. To join two independent clauses, use the conjunction **as, as if**, or **as though** instead:

    ✗   The transformer looked <u>like</u> it was overloaded.

    ✓   The transformer looked <u>as though</u> it was overloaded.

    ✓   The residue looked <u>like</u> sand.

loose, lose. Loose, as an adjective, means the opposite of tight. Its verb form is **loosen. Lose** as a verb, means misplace or be defeated. Watch the spelling.

    ✗   The battery was <u>loosing</u> its charge.

    ✓   The battery was <u>losing</u> its charge.

media, medium. Use **media** as the plural of the singular word **medium**:

    The <u>media</u> have reported the increasing popularity of polyurethane foam as an insulation <u>medium</u>.

minimize. See **diminish**.

mitigate, militate. To **mitigate** means *to reduce the severity of* something; to **militate against** something means *to oppose* it.

myself, me. Myself is an intensifier of (not a substitute for) *I* or *me*. If you use it, include *I* or *me* in your sentence first:

    ✗   Please contact John or <u>myself</u> if you have questions.

    ✓   Please contact John or <u>me</u>.

    ✗   Jane and <u>myself</u> are presenting our findings.

    ✓   Jane and <u>I</u> are presenting our findings

    ✓   I completed the research <u>myself</u>.

number, amount. See **amount, number**.

off of. Drop the redundant **of**:

> ✗ The fence kept children <u>off of</u> the premises.
>
> ✓ The fence kept children <u>off</u> the premises.

orient, orientate. As a verb, **orient** means *to position something properly, to determine its bearings*. Although **orientate** is used with the same meaning, the shorter word is preferred.

passed, past. **Passed** is the past form of the verb **pass**. **Past** refers either to time or to one place farther than another:

> <u>Past</u> the bridge, they <u>passed</u> a sports car.

people, persons. Legal contexts (for example, elevator licences) use **persons** as the plural of **person** in case of liability. For all other purposes, use **people**:

> ✗ The media reported ten <u>persons</u> missing after the ferry sank.
>
> ✓ Twenty <u>people</u> registered for the first-aid class.

per cent, percentage. **Per cent** (from the Latin *per cent*) is used for numbers written as words (*four per cent*); with digits, use the symbol instead (*4.0%*). **Percentage** means proportion or amount and is used in generalizing about measurements:

> What <u>percentage</u> of students graduate with honours?
>
> Twenty-five <u>per cent</u> of students graduate with averages over 80<u>%</u>.

phenomenon. **Phenomenon** is a singular noun; the plural is **phenomena**. Not to be used as a synonym for *occurrence*, **phenomenon** refers to an exceptional circumstance.

practice, practise. **Practice** can be a noun or a modifier; **practise** is always a verb. (Note, however, that the standard American spelling is **practice** for both.)

> In a <u>practice</u> game, players <u>practise</u> their skills.

precede, proceed. To **precede** is *to go before* (earlier) *or in front of* others; to **proceed** is *to go on* or *to go ahead*:

> The dean's welcome preceded the awarding of certificates.

> The presentation proceeded without interruption.

prescribe, proscribe. These words have quite different meanings. **Prescribe** means *to advise the use of* or *to impose authoritatively*. **Proscribe** means *to reject, denounce,* or *ban*:

> The professor prescribed the conditions for the experiment.

> The student government proscribed the publication of unsigned editorials on its blog.

preventive, preventative. Although both are used to describe measures of prevention, the shorter word, **preventive**, is preferred.

principle, principal. **Principle** is a noun meaning *a general truth or law*; **principal** can be used as either a noun or an adjective, meaning *chief*.

> Her principal error lay in not understanding the principle.

rational, rationale. **Rational** is an adjective meaning *logical* or *able to reason*. **Rationale** is a noun meaning *explanation*:

> That was not a rational decision.

> The president's memo explained the rationale for the decision.

real, really. **Real**, an adjective, means *true* or *genuine*; **really**, an adverb, means *actually, truly, very,* or *extremely*:

> The nugget was of real gold.

> The gold nugget was really valuable.

remanent, remnant. **Remanent** is an adjective for describing residual magnetization (or *remanence*). **Remnant** is a noun naming something left over or remaining.

sceptic, septic. British and Canadian spelling prefers **sceptic** for a person who doubts or disbelieves (the American spelling is *skeptic*). Be careful not to confuse it with the lookalike **septic**, which is used with ulcers and sewers.

seasonable, seasonal. **Seasonable** means usual or suitable for the season; **seasonal** means of, depending on, or varying with the season:

> Meteorologists predict the return of seasonable temperatures later in the week.

> In the summer, employment figures are adjusted to account for seasonal employment increases.

simple, simplistic. Both words refer to something that is *easy* or *basic*. **Simplistic**, however, has the negative connotation of *oversimplification*, so it should not be used as a synonym.

> The solution was simple.

> His simplistic answer annoyed the examiners.

stimulant, stimulus. A **stimulant** temporarily increases some vital process in an organ or organism, where a **stimulus** incites the organism to act in the first place.

than, then. **Than** is a conjunction linking unequal comparisons; **then** is an adverb of time or sequence:

> A is shorter than B.

> Shorten A first then B.

that, which. **That** can introduce a restrictive clause, and **which** can introduce a clause that is either restrictive or non-restrictive (see pp. 220–221, 259).

> They studied cybernetics, which is a modern academic domain that touches all traditional disciplines.

Grammar-checkers point to an error when there is no comma before **which**, so it is safe to make the distinction:

> He followed the format that his company requires.

> He followed IEEE format, which his company requires.

their, there. **Their** shows possession for the third-person plural pronoun. **There** is usually an adverb, meaning *at that place* or *at that point*:

> They parked their bikes there.

> There is no point arguing.

tortuous, torturous. The adjective **tortuous** means *full of twists and turns* or *circuitous*. **Torturous**, derived from *torture*, means *involving torture* or *excruciating*:

> The graph presented a tortuous curve.

> The interview was a torturous experience for the applicant.

translucent, transparent. A **translucent** substance permits light to pass through, but not enough for a person to see through it; a **transparent** substance permits light to pass unobstructed, so that objects can be seen clearly.

try and. This is a feature of conversational English. Use **try to** instead.

> ✗   They were going to try and visit the exhibition.

> ✓   They were going to try to visit the exhibition.

turbid, turgid. **Turbid**, with respect to a liquid or colour, means *muddy, not clear,* or (with respect to writing style) *confused*. **Turgid** means *swollen, inflated,* or *enlarged,* or (again with reference to style) *pompous* or *bombastic*.

unique. This word, which literally means *of which there is only one,* is both overused and misused. As a synonym of *unparalleled* or *unequalled,* it should not be used in comparisons such as *more unique* or *very unique*.

use, usage. Distinguish between these seeming synonyms by employing **usage** only when you mean *usual practice* or *proper* **use**:

> Consult the manual about the usage of this device.

> Use of this device will save time.

Note that many caution against the wordiness of *the use of*. Often, you can save words by omitting the phrase:

> [The use of] calculators will not be permitted during the exam.

while. To prevent misreading, use **while** only when you mean *at the same time that*. Avoid using **while** as a substitute for *although, whereas,* or *but*:

- ✗ While she's getting fair marks, she'd like to do better.
- ✓ He headed for home, while she decided to stay.
- ✓ He fell asleep while he was reading.

-wise. Avoid using **-wise** as a suffix to form new words that mean *with regard to*, for the tone is too casual for formal writing.

- ✗ Saleswise, the company did better last year.
- ✓ The company's sales have decreased this year.

your, you're. **Your** is a possessive pronoun; **you're** is a contraction of *you are*:

> You're likely to miss your train.

Chapter Checklist

- ☐ Remember that a spellchecker isn't always a helpful proofreader. Watch out for homophones—words that sound the same but are spelled differently. Your spellchecker doesn't distinguish between these.
- ☐ Keep a list of words that have confused you and watch for them in particular as you proofread.
- ☐ Book mark reliable websites like Purdue's Online Writing Lab or the Government of Canada's language portal for a quick link to resources.

Glossary

abbreviation. A short form of a word, often consisting of the word's initial letters and a period; *eng.* is a typical abbreviation of *engineering*.

abscissa. Another name for the horizontal or x-axis on a line graph.

absolute value. The magnitude of a number, often calculated by taking the square root of the square of a number.

abstract. A comprehensive summary (50–250 words) accompanying a formal report, proposal, or paper and outlining its contents. As an adjective, **abstract** describes something theoretical or intangible rather than concrete.

acronym. A pronounceable word made up of the first letters of the words in a phrase or name: e.g., *SAW* (*surface acoustic wave*). A group of initial letters that are pronounced separately is an **initialism**: e.g., *GPS, UHF*.

active voice. See **voice**.

adjective. A word that modifies or describes a noun or pronoun: e.g., *cloudy, new, legal*. An **adjective phrase** or **clause** is a group of words modifying a noun or pronoun: e.g., *the results that are expected*.

adverb. A word that modifies or qualifies a verb, adjective, or adverb, often answering a question such as *how? why? when?* or *where?*: e.g., *slowly, absolutely, soon, there*. An **adverb phrase** or **clause** is a group of words functioning like an adverb: e.g., *in the lab, if X is valid*. (See also **transitional adverb**.)

agreement. Consistency in tense, number, or person between related parts in a sentence: e.g., between subject and verb, or noun and related pronouns (see pp. 182–185).

algorithm. A step-by-step approach to doing something specific; this term is most often used to describe computer procedures.

ambiguity. Vague or equivocal language with meaning that can be taken two or more ways.

analysis. A structured approach to thinking, or the results of thinking in a structured way.

antecedent (or **referent**). The noun for which a following pronoun stands: e.g., *engineers* in *Engineers are experts in their fields*.

appendix. A section added at the end of a work to include material not essential to the main discussion but complementary to it.

appositive. A word or phrase that identifies a preceding noun or pronoun: e.g., *Dr. Lee, my supervisor, is a fine mentor*. The second phrase (*my supervisor*) is said to be **in apposition to** the first (*Dr. Lee*).

article. A word introducing a noun that identifies whether the noun is general or specific, singular or plural. *A* and *an* are called **indefinite articles**; *the* is a **definite article**. (See also **determiner**.)

assertion. A positive statement or claim: e.g., *The data are convincing*.

auxiliary verb. A verb forming the tenses, moods, and voices of other verbs: e.g., *was* in *It was done*. The main auxiliary verbs in English are *be, have,* and *do. Can, could, may, might, must, shall, should, will,* and *would* are called **modal auxiliaries**.

average (mean), median, mode. Three measures used to characterize the centre point of a set of values. The **average** (or **mean**) is calculated by adding a set of values and then dividing the result by the number of values added. **Median** represents the middle number in a set of numbers. **Mode** is the most frequently occurring value.

bibliography. 1. A list of works used or referred to in writing a paper or report. 2. A reference book listing works available on a particular subject.

calibration. The process of preparing a measuring instrument for use, either by calculating an error factor or by eliminating a difference between the expected value and the measured value for some standard.

case. Any of the inflected forms of a personal pronoun (see **inflection**).
  Subjective case: *I, we, you, he, she, it, they, who.*
  Objective case: *me, us, you, him, her, it, them, whom.*
  Possessive adjective case: *my, our, your, his, her, its, their, whose.*
  Possessive noun case: *mine, ours, yours, his, hers, its, theirs, whose.*

circumlocution. A roundabout or circuitous expression, often used in a deliberate attempt to be vague or evasive: e.g., using *at this particular point in time* for *now.*

clause. A group of words containing a subject and a predicate. An **independent clause** can stand by itself as a complete sentence: e.g., *The solution remained stable.* A **subordinate** (or **dependent**) **clause** cannot stand by itself but must be connected to an **independent clause**: e.g., *Unless the temperature rises, the solution remains stable.*

cliché. A phrase or idea that has lost its impact through overuse: e.g., *beyond the shadow of a doubt; window of opportunity.*

collective noun. A noun that is singular in form but refers to a group: e.g., *family; team; staff.* It will take a verb that is either singular or plural, depending on whether it refers

to the group as a whole or its individual members.

colloquial language. Everyday expressions that are appropriate to speaking. Formal writing avoids **colloquialisms**: e.g., *We've got to get the bugs out of the program.* [Revised: *We must eliminate the programming errors.*]

comma splice. See **run-on sentence**.

complement. A completing word or phrase that usually follows a linking verb to form a **subjective complement**: e.g., (1) *She is my colleague.* (2) *That substance tastes bitter.* When the complement is an adjective, it can be called a **predicate adjective**. An **objective complement** completes the direct object rather than the subject: e.g., *We considered the assessment fair.*

complex sentence. A sentence containing at least one dependent clause in addition to an independent one: e.g., *The estimates were accepted although they were sketchy.*

compound sentence. A sentence containing two or more independent clauses: e.g., *Several solutions exist, but only one is viable.* A sentence is called **compound-complex** if it contains a dependent clause as well as two independent clauses: e.g., *Several solutions exist, but only one is viable if there are budgetary constraints.*

conclusion. The part of a paper or report in which the findings are pulled together or the implications revealed so that the reader has a sense of closure or completion.

concrete language. Specific language that provides descriptive details: e.g., *grey crystalline deposits, rusted barbed wire, toxic waste.*

conditional verb. The verb form called for in hypothetical situations: e.g., *If the data were recalculated, the results would be different* (see pp. 190–191).

confidence interval. A statistical calculation that yields a range of values instead of a single estimate for some aspect of a population.

conjunction. A word used to link words, phrases, or clauses. A **coordinating conjunction** (*and, nor, but, or, yet*) links two equal (parallel) parts of a sentence (see **correlative conjunctions**). A **subordinating conjunction**, placed at the beginning of a subordinate clause, shows the logical dependence of that clause on another: e.g., (1) *Although the work is complete, it is unsatisfactory.* (2) *They ran several tests after they installed the software.* Do not confuse conjunctions with transitional (conjunctive) adverbs.

connotation. The range of ideas or meanings suggested by a certain word in addition to its literal meaning. Apparent synonyms, such as *artificial* and *imitation* and *synthetic*, even *counterfeit*, have differing connotations. (See **denotation**.)

context. The setting, background, or foundation for an investigation that helps establish its place in the larger scheme of things.

contraction. A word formed by combining and shortening two or more words: e.g., *isn't* from *is not*; *they'll* from *they will*. Contractions are appropriate in letters and casual messages, not in formal reports or academic writing.

coordinating conjunction. See **conjunction**.

copula verb. See **linking verb**.

copyright. A legal mechanism used by authors, illustrators, mapmakers, software developers, and others to keep people from using their work without permission.

correlative conjunctions. Pairs of coordinating conjunctions that call for **parallel wording**: e.g., *either/or; neither/nor; not/but; not only/but (also)*. (See p. 178 and p. 201.)

Critical Path Method. A project management method that models the activities and events of a project to determine how much time is needed to complete the project. (See p. 133.)

culture. The values and beliefs shared by a group of people, e.g., members of a profession, organization, or company.

dangling modifier. A modifying word or phrase (often including a participle) that is not grammatically connected to any part of the sentence: e.g., *Examining the results, several discrepancies were noted.* (See p. 198.)

deduction. A research process that tests a *hypothesis* and draws conclusions based on test results. (See pp. 5–6.)

demonstrative pronoun. A pronoun that points to something: e.g., (1) *This is the result*; (2) *That looks like the reason.* When used to modify a noun or pronoun, a demonstrative becomes a kind of **adjective**: *this reaction; those data.*

denotation. The literal or dictionary meaning of a word. (See **connotation**.)

determiner. A word introducing a noun that identifies whether the noun is general or specific, countable or uncountable, singular or plural. All singular countable nouns must be introduced by a determiner. (See also **article**.)

diction. The choice of words with respect to their tone, degree of formality, or register. Formal diction is the language of research papers, reports, and legal contracts. Public presentations and business correspondence call for less formal diction, but they are not as informal as the diction of everyday speech or conversational writing, which often includes slang.

digital object modifier (DOI). A unique alphanumeric string, assigned to works found online, that contains a prefix and a suffix separated by a slash (for example, 10.1037/0096-3445). Every article with a DOI is registered and can be easily identified and retrieved electronically, no matter whether the website where you found the article continues to exist.

discourse. Talk, either oral or written. **Direct discourse** (or **direct speech**) cites an actual quotation in its context: e.g., *Edison said, "There is no substitute for hard work."* In writing, direct discourse is put in quotation marks. **Indirect discourse** (or indirect speech) gives the gist of the quotation rather than a word-for-word citation. Quotation marks are not used, although the reference is documented: e.g., *Edison said that nothing could replace hard work.*

ellipsis marks. Three spaced dots indicating an omission from a quoted passage [. . .].

empirical evidence. Data created by making measurements or doing experiments.

encryption. A process to convert information into code so that it cannot be read by someone who has not been given access, most often to protect information sent over the Internet.

endnote. A citation or note appearing at the end of a paper or a chapter in a book.

entropy. The tendency of all physical systems to move toward disorder.

e-portfolio. A digital collection of artifacts (such as writings, art, videos, slideshows) that represent an individual's volunteer, employment, curricular and extra-curricular activities and encourage reflection on the learning they exemplify.

equilibrium. A state in which opposing forces or actions are in balance.

error bar. A graphical depiction of a confidence interval.

error factor. The difference between the actual value and the measured value as calculated during **calibration**. This factor is used to adjust all subsequent measurements to account for the systematic error created by the measuring instrument.

essay. A composition on any subject. Some essays are descriptive or narrative; in an academic setting, most are expository (explanatory) or persuasive.

euphemism. A word or phrase used to avoid some other wording that might be considered offensive or harsh: e.g., *custodian* for *janitor*; *underperform* for *fail*.

expletive. 1. A word or phrase used to fill out a sentence without adding to the sense: e.g., <u>To be sure</u>, it's not an ideal situation. 2. A swear word.

exploratory writing. Informal writing done to help generate ideas before formal planning begins.

fair dealing. The limited right of researchers and others to borrow small amounts of a copyrighted work without official permission, yet with documentation.

fault trees. A structured process used to identify and assess potential causes of system failure before the failures actually occur. Analysts begin by selecting a top-level event, like a critical safety issue, and then work down to evaluate all the contributing events that may ultimately lead to the occurrence of that top-level event. The resulting fault tree diagram is a graphical representation of the chain of events in the system.

fission. A reaction in which the atomic nucleus is split and energy is released. This reaction is the key process within both the atomic bomb and nuclear power generators.

flowchart. A graphical representation of a process. (See p. 14.)

footnote. A citation or note appearing at the bottom of the page in a paper or article.

fused sentence. A sentence combining two independent clauses without punctuation. (See **run-on sentence**.)

fusion. A reaction in which several atomic nuclei combine and energy is released. This process takes place continuously in the sun and stars.

Gantt chart. A system of coordinating multiple tasks along a timeline that allows managers to visualize all related activities at once. (See p. 63.)

gerund. A **verbal** marked by its *-ing* ending and functioning as a noun in a sentence: e.g., *We look forward to interviewing the candidate.*

grammar. The study of the forms and relations of words and of the rules governing their use in speech and writing.

Greek alphabet. The alphabet used to represent variables and constants in equations. By convention, some of the characters are regularly used to represent a specific variable or constant: e.g., *pi* (π) represents the commonly used irrational number 3.1415+.

homophone. A term meaning "same sound" and used to describe words that are pronounced the same but spelled differently (than, then; too, to, two; etc.), also referred to by the term *homonym* ("same name").

hypothesis. A testable statement that indicates what an experimenter expects to find through his or her work.

hypothetical instance. An imagined occurrence, generally indicated by a clause beginning with *if.* (See **conditional verb.**)

idiom. A phrase that has no explanation for its structure other than that it "sounds right": e.g., *This makes sense.* (See p. 208.)

independent clause. See **clause.**

indirect discourse (or **indirect speech**). See **discourse.**

induction. A research process in which investigators come to a conclusion based on what they have observed. (See pp. 5–6.)

inertia. The tendency of a body to resist acceleration.

infinitive. A **verbal** made up of the root form of a verb usually introduced by the marker *to*: e.g., *to evaluate.* The base infinitive omits the *to*: e.g., *evaluate.*

inflection. The change in form of a word to indicate number, person, case, tense, or degree: e.g., *Athletes run fast. That athlete ran faster.*

initialism. See **acronym.**

intensifier (or **qualifier**). A word that modifies and adds emphasis to another word or phrase: e.g., *very slowly; quite important.* Avoid adding such words to absolutes like *major* or *empty* or *complete.*

interjection. An expression or exclamation, usually followed by a comma or exclamation mark: e.g., *Well, that's important. Ouch!*

interrogative sentence. A sentence that asks a question and that concludes with a question mark: e.g., *How much time elapsed? Did the link work?*

interrogative words. Question words used in direct queries or in noun clauses: e.g., *what, who, how,* etc.

intransitive verb. A verb that does not take a direct object: e.g., *fall, sleep, happen.*

introduction. A section at the start of a report or paper that tells the reader what will be discussed—and why.

italics. Slanting type represented in handwriting or typescript by underlining.

jargon. Technical terms when used unnecessarily or inappropriately: e.g., *resource recovery facility* for *incinerator.* (See p. 162.)

linking verb (or **copula verb**). A verb such as *be, seem,* or *feel* used to join subject to complement: e.g., *The strategy is complex.*

literal meaning. The primary, or denotative, meaning of a word.

logarithm. The power to which a base must be raised to produce a given number. The common logarithm's base is 10, the natural logarithm's base is *e*, and the binary logarithm's base is 2. For example, the common logarithm of 100 is 2, the natural logarithm of *e* is 1, and the binary logarithm of 8 is 3.

matrix. The mathematical representation in which data and equations are organized into rectangular arrays of numbers and variables.

mean, median, mode. See **average**.

misplaced modifier. A word or group of words that can cause confusion by not being placed next to the element it modifies: e.g., *We nearly reviewed twenty sources*. [Revised: *We reviewed nearly twenty sources*.] (See p. 197.)

modal auxiliary. See **auxiliary**.

modifier. A word or group of words that describes or limits the context of another element in the sentence.

mood. 1. As a grammatical term, the form that shows a verb's purpose:
 Indicative mood: *She is working hard*. [a statement]
 Imperative mood: *Work hard!* [a command]
 Interrogative mood: *Is she working hard?* [a question]
 Subjunctive mood: *It is essential that she work hard*. [a condition]
 2. When applied to literature generally, the atmosphere or tone created by the author.

negatives. Words or constructions that carry a negative meaning: e.g., *no, not, never, none, nobody, nothing*.

non-restrictive modifier. See **restrictive modifier**.

noun. An inflected part of speech naming a person, place, thing, idea, or feeling—and usually serving as subject, object, or complement. A **common noun** is a general term: e.g., *day, school, automobile*. A **proper noun** is a specific name signalled by an introductory capital letter: e.g., *Wednesday, Queen's University*, Toyota.

number. Generally, a distinction between singular (1) and plural ($>$1). **Cardinal numbers** represent how many of something there are (*one, two, three*, etc.), while **ordinal numbers** represent a ranking or position (*first, second, third*, etc.)

object. 1. A noun or pronoun (called the **direct object**) that completes the action of the verb: e.g., *He asked a question*. An **indirect object** is the person or thing receiving the direct object: e.g., *He asked her* [indirect object] *a question* [direct object]. 2. The noun or pronoun in a group of words beginning with a preposition: e.g., *between them*; *in the corner*.

objective complement. See **complement**.

objectivity. A position or stance taken without bias or prejudice. (Compare **subjectivity**.)

optimization. Mathematical techniques used to make a decision, a system, or a design as effective or as functional as possible.

ordinate. Another name for the vertical or y-axis on a line graph.

outline. With regard to a paper, presentation, or report, a brief sketch of the main parts, to help in organizing them; a written plan.

paragraph. A set of sentences arranged logically to explain or describe an idea, event, or object; the start of a paragraph is marked either by an indentation or by double spacing.

parallel wording. Wording in which a series of items has a similar grammatical form:

e.g., *The Babylonians, the Egyptians, and even the Bible note the existence of pi.* (See p. 199.)

parameters. Constants whose values characterize aspects of a system when the system's behaviour is represented as a set of equations.

paraphrase. A restatement of a sentence or passage using different words from the original.

parentheses. Curved lines enclosing and setting off a passage ( ); not to be confused with square brackets [ ].

parenthetical element. A word or phrase inserted as an explanation or afterthought into a passage that is grammatically complete without it: e.g., *Fuzzy logic, in more than one sense, resembles human decision making.*

participle. A verbal that functions as a modifier. Participles can be either **present**, marked by an *-ing* ending (e.g., *being* or *having* or *taking*), or **past** (e.g., *taken*); they also exist in the **passive**: e.g., *being taken* or *having been taken*.

part of speech. Each of the major categories into which words are sorted according to grammatical function. Some grammarians consider only inflected words (nouns, verbs, adjectives, and adverbs); others include prepositions, pronouns, conjunctions, and interjections.

passive voice. See **voice**.

past participle. See **participle**.

patent. A legal mechanism by which a person or company who owns a patent can sue others who make, use, or sell the product or process without the permission of the patent owner.

periodic sentence. A sentence in which the normal order is inverted or in which an essential element is suspended until the very end: e.g., *Not until the early twentieth century, with the development of reinforced concrete, was it possible to build a skyscraper.*

person. In grammar, the three classes of personal pronouns referring to the person speaking (**first person**: *I, we*), the person(s) spoken to (**second person**: *you*), and the distant person(s) spoken about (**third person**: *he, she, it, they*).

personal pronouns. See **pronoun**.

phrase. A set of words lacking a subject–predicate combination, typically forming part of a clause. The most common type is the **prepositional phrase**—a unit consisting of a preposition and an object: e.g., *Experiments were divided into three stages.*

plagiarism. The deliberate use of someone else's work or ideas without acknowledgement.

plural. Indicating two or more, specifically with numbers. Nouns, pronouns, and verbs all have plural forms.

possessive case. See **case**.

precision. The number of significant digits to which a value has been measured.

prefix. An element placed in front of the root form of a word to make a new word: e.g., *pro-, in-, sub-, anti-*. (Compare **suffix**.)

preposition. A short word or phrase heading a group of words representing an object; together, a preposition and its object make up a **prepositional phrase**: e.g., *under the time limit; because of the constraints.*

process. A series of actions or operations leading to a particular and usually desirable result.

pronoun. A word that stands in as a noun. A personal pronoun stands for the name of a person: *I, you, she, he, it, we, they.* (see **person**.)

public domain. A term applied to work that lacks copyright protection; permission for copying is not required, although everything should still be documented.

punctuation. A conventional system of signs and symbols (e.g., comma, period, semicolon) used to indicate stops or divisions in a sentence.

qualitative data. Data that can be satisfactorily described not by numbers but by words.

quantitative data. Data that can be measured or enumerated.

quotation. The recording or repeating of something someone has said or written.

random error. Variations observed in a data set but not explicable in a study, also a bid or estimate of the costs of a project.

range. A measure of the variability of scores in a sample. It tells how closely a set of scores cluster around the mean, and it is equal to the difference between the highest and lowest observed values in a data set.

redundancy. Unnecessary or ineffectual repetition: e.g., *join together*; *first and foremost*. (See pp. 163–168.)

reference works. Sources consulted when preparing a paper or report.

referent. See **antecedent**.

reflexive (intensive) pronoun. A pronoun ending with *-self* or *-selves* to echo or emphasize a preceding noun or pronoun: e.g., *The researchers congratulated themselves.*

register. Levels of formality (from high to low) in word choice and sentence structure.

relative clause. A clause introduced by a relative pronoun: e.g., *The approach that was taken was the most cost effective.*

relative pronoun. *Who, which, that, whose,* or *whom,* used to introduce an adjective clause: e.g., *the approach that was taken.*

request for proposal (RFP). A formal invitation describing project requirements. The appropriate response to an RFP is a formal proposal outlining both how the respondent plans to meet the project requirements and the expected compensation for completing the project.

restrictive modifier. A phrase or clause that identifies or is essential to the meaning of a term: e.g., *They read the article that their professor had recommended.* It should not be set off by commas. A **non-restrictive modifier** is not needed for identification and is usually set off parenthetically with commas: e.g., *They read Jennings' article, which their professor had recommended.* (See pp. 230-231.)

rhetorical question. A question posed and answered by a writer or speaker to draw attention to a point; no response is expected on the part of the audience: e.g., *How important are these findings? Indeed, they are significant for several reasons.*

risk management. Actions taken to understand, control, and limit the possibility of adverse consequences.

run-on sentence. A sentence that goes beyond the point where it should have stopped. The term covers both the **comma splice** (two sentences incorrectly joined by a comma) and the **fused sentence** (two sentences incorrectly joined without any punctuation). (See pp. 183–184.)

scientific method. A structured approach to research that requires the formulation of a hypothesis based on a systematic and objective collection and review of existing data followed by the experimental testing of that hypothesis.

scientific notation. A convention in which a real number is expressed as the product of a real number and a base (usually 10) raised to a power: e.g., $3.257 \times 3\ 10^5$ represents 325,700.

sentence. A grammatical unit that includes both a subject and a predicate. The end of a sentence is marked by a period.

sentence fragment. A group of words lacking either a subject or a complete predicate; an incomplete sentence punctuated as a sentence. (See pp. 181–182.)

significant digits. The number of decimal places to which a result can be meaningfully reported when the result is expressed according to scientific notation: e.g., 3.257×10^5 has four significant digits.

simple sentence. A sentence made up of only one clause: e.g., *Data fusion is a popular information processing technique.*

slang. Colloquial speech inappropriate for academic or professional writing, often used in a special sense by a limited group: e.g., *clicks* for *kilometres per hour*; *ace* for *succeed.*

squinting modifier. A kind of misplaced modifier that could be connected to elements on either side, producing ambiguity: e.g., *Students who study often do well on exams.* (See p. 197.)

standard deviation. A measure of the variability of scores in a sample, showing the average distance of a set of scores from the mean.

standard English. The English currently spoken or written by literate people and widely accepted as the correct and conventional form.

statistics. A term referring to both (1) the set of techniques for estimating the characteristics of a population based on the characteristics of a sample drawn from that population, and (2) the estimates created using these techniques.

stochastic process. A variable that changes in a way that depends, at least partially, on chance. Such variables are measured and studied using probability and statistics.

subject. In grammar, the noun or noun equivalent with which the verb agrees and about which the rest of the clause is predicated:

e.g., *Research is a challenge, but many people thrive on it.*

subjective complement. See **complement**.

subjectivity. A personal stance that is based on feelings or opinions and is not impartial or disinterested. (Compare **objectivity**.)

subjunctive. See **conditional verb** and **mood**.

suboptimization. Optimization of a subsystem. The principle states that suboptimization does not always lead to global optimization.

subordinate clause. See **clause**.

subordinating conjunction. See **conjunction**.

subordination. Making one clause dependent upon another (see **clause**).

suffix. An element added to the end of a word to form a derivative: e.g., *-tion, -ify, -ment, -ize, -ful* . (Compare **prefix**.)

synonym. A word with the same dictionary meaning as another word: e.g., *begin* and *start*.

syntax. Sentence construction; the grammatical relationship of words and phrases.

synthesis. A process in which separate elements combine to create a coherent whole.

tense. A set of forms, often inflected, that are taken by a verb to indicate past, present, or future time (i.e., they *went*, it *goes*, we *will go*).

theme. A recurring or dominant idea.

theory. A set of general statements or abstract principles created to explain a set of facts or phenomena.

thesaurus. A dictionary grouping words with their synonyms and antonyms, used for locating a word with the most appropriate connotation.

thesis statement. A one-sentence assertion that presents the central argument of a paper.

topic sentence. The sentence in a paragraph that expresses the main or controlling idea.

transitional adverb. A word or phrase that shows the logical relation between sentences or parts of a sentence and thus helps to signal the change from one idea to another: e.g., *therefore, however, for example*.

transitive verb. A verb that takes an object: e.g., *heat, bring, finalize*. (Compare **intransitive verb**.)

transitivity. An aspect of a relationship between objects. Equality is **transitive** because $a = b$ and $b = c$ implies that $a = c$.

true value. The real value that researchers strive to obtain when measuring a characteristic of a system or a descriptor of a population.

usage. The way in which a word or phrase is normally and correctly used; accepted practice.

variable. An element that can represent any one of a set of values. In an experiment, the **independent variable** is the one manipulated by the experimenter; the **dependent variable** is what the experimenter predicts will be affected by manipulations of the independent variable.

variance. A measure of the variability of scores in a sample, telling how closely a set of scores cluster around the mean.

verb. That part of the predicate expressing an action, process, state, or condition, telling what the subject is or does. Verbs are inflected to show **tense** (time). The principal parts of a verb are the basic forms from which all tenses are made: the **infinitive**, the **present tense,** the **past tense**, the **present participle** and the **past participle**.

verbal. A word that resembles a verb in form but does not function as one: a **participle**, a **gerund**, or an **infinitive**.

voice. The form of a verb that shows whether the subject acted (**active voice**) or was acted upon (**passive voice**): e.g., *They presented the award* (active). *The award was presented by them* (passive). Only **transitive** verbs (verbs taking objects) can be made passive.

Appendix

Weights, Measures, and Notation

The conversion factors are not exact unless so marked. They are given only to the accuracy likely to be needed in everyday calculations.

1. IMPERIAL AND AMERICAN, WITH METRIC EQUIVALENTS

Linear measure

1 inch	= 25.4 millimetres exactly
1 foot = 12 inches	= 0.3048 metre exactly
1 yard = 3 feet	= 0.9144 metre exactly
1 (statute) mile = 1,760 yards	= 1.609 kilometres (km)
1 int. nautical mile = 1.150779 miles	= 1.852 km exactly

Square measure

1 square inch	= 6.45 sq. centimetres
1 square foot = 144 sq. in.	= 9.29 sq. decimetres
1 square yard = 9 sq. ft.	= 0.836 sq. metre
1 acre = 4,840 sq. yd.	= 0.405 hectare
1 square mile = 640 acres	= 259 hectares

Cubic measure

1 cubic inch	= 16.4 cu. centimetres
1 cubic foot = 1,728 cu. in.	= 0.0283 cu. metre
1 cubic yard = 27 cu. ft.	= 0.765 cu. metre

Capacity measure

Name	System	Equal to	Metric
fluid oz.	imperial	1/20 imp. pint	28.41 ml
	US (liquid)	1/16 US pint	29.57 ml
gill	imperial	1/4 pint	142.07 ml
	US (liquid)	1/4 pint	118.29 ml
pint	imperial	20 fl.oz.(imp.)	568.26 ml
	US (liquid)	16 fl.oz.(US)	473.18 ml
	US (dry)	1/2 quart	550.61 ml
quart	imperial	2 pints	1.1365 litres
	US (liquid)	2 pints	0.9464 litre
	US (dry)	2 pints	1.1012 litres
gallon	imperial	4 quarts	4.546 litres
	US (liquid)	4 quarts	3.785 litres
peck	imperial	2 gallons	9.092 litres
	US (dry)	8 quarts	8.810 litres
bushel	imperial	4 pecks	36.369 litres
	US (dry)	4 pecks	35.239 litres

Avoirdupois weight

1 grain	= 0.065 gram
1 dram	= 1.772 grams
1 ounce = 16 drams	= 28.35 grams
1 pound = 16 ounces = 7,000 grains	= 0.45359237 kilogram exactly
1 stone = 14 pounds	= 6.35 kilograms
1 quarter = 2 stones	= 12.70 kilograms
1 hundredweight = 4 quarters = 112 lb.	= 50.80 kilograms
1 (long) ton = 20 cwt. = 2,240 lb.	= 1.016 tonnes
1 short ton = 2,000 pounds	= 0.907 tonne

2. METRIC, WITH IMPERIAL EQUIVALENTS

Linear measure

1 millimetre	= 0.039 inch
1 centimetre = 10 mm	= 0.394 inch
1 decimetre = 10 cm	= 3.94 inches
1 metre = 100 cm	= 1.094 yards
1 decametre = 10 m	= 10.94 yards
1 hectometre = 100 m	= 109.4 yards
1 kilometre = 1,000 m	= 0.6214 mile

Square measure

1 square centimetre	= 0.155 sq. inch
1 square metre = 10,000 sq. cm	= 1.196 sq. yards
1 are = 100 sq. metres	= 119.6 sq. yards
1 hectare = 100 ares	= 2.471 acres
1 square kilometre = 100 ha	= 0.386 sq. mile

Cubic measure

1 cubic centimetre	= 0.061 cu. inch
1 cubic metre = one million cu. cm	= 1.308 cu. yards

Capacity measure

1 millilitre	= 0.002 pint (imperial)
1 centilitre = 10 ml	= 0.018 pint
1 decilitre = 100 ml	= 0.176 pint
1 litre = 1,000 ml	= 1.76 pints
1 decalitre = 10 l	= 2.20 gallons (imperial)
1 hectolitre = 100 l	= 2.75 bushels (imperial)

Weight

1 milligram	= 0.015 grain
1 centigram = 10 mg	= 0.154 grain
1 decigram = 100 mg	= 1.543 grain
1 gram = 1,000 mg	= 15.43 grain
1 decagram = 10 g	= 5.64 drams
1 hectogram = 100 g	= 3.527 ounces
1 kilogram = 1,000 g	= 2.205 pounds
1 tonne (metric ton) = 1,000 kg	= 0.984 (long) ton

3. SI UNITS
Base units

Physical quantity	Name	Abbr. or symbol
length	metre	m
mass	kilogram	kg
time	second	s
electric current	ampere	A
temperature	kelvin	K
amount of substance	mole	mol
luminous intensity	candela	cd

Supplementary units

Physical quantity	Name	Abbr. or symbol
plane angle	radian	rad
solid angle	steradian	sr

Derived units with special names

Physical quantity	Name	Abbr. or symbol
frequency	hertz	Hz
energy	joule	J
force	newton	N
power	watt	W
pressure	pascal	Pa
electric charge	coulomb	C
electromotive force	volt	V
electric resistance	ohm	Ω
electric conductance	siemens	S
electric capacitance	farad	F
magnetic flux	weber	Wb
inductance	henry	H
magnetic flux density	tesla	T
luminous flux	lumen	lm
illumination	lux	lx

4. TEMPERATURE

Celsius (or Centigrade): Water boils (under standard conditions) at 100° and freezes at 0°

Fahrenheit: Water boils at 212° and freezes at 32°

Kelvin: Water boils at 373.15 kelvins and freezes at 273.15 kelvins

Celsius	Fahrenheit
−17.8°	0°
−10°	14°
0°	32°
10°	50°

Celsius	Fahrenheit
20°	68°
30°	86°
40°	104°
50°	122°
60°	140°
70°	158°
80°	176°
90°	194°
100°	212°

To convert Celsius into Fahrenheit: multiply by 9, divide by 5, and add 32.

To convert Fahrenheit to Celsius: subtract 32, multiply by 5, and divide by 9.

5. METRIC PREFIXES

	Abbr. or symbol	Factor
deca-	da	10
hecto-	h	10^2
kilo-	k	10^3
mega-	M	10^6
giga-	G	10^9
tera-	T	10^{12}
peta-	P	10^{15}
exa-	E	10^{18}
deci-	d	10^{-1}
centi-	c	10^{-2}
milli-	m	10^{-3}
micro-	μ	10^{-6}
nano-	n	10^{-9}
pico-	p	10^{-12}
femto-	f	10^{-15}
atto-	a	10^{-18}

These prefixes may be applied to any units of the metric system: hectogram (abbr. hg) = 100 grams; kilowatt (abbr. kW) = 1,000 watts; megahertz (MHz) = 1 million hertz; centimetre (cm) = ¹⁄₁₀₀ metre; microvolt (μV) = one-millionth of a volt; picofarad (pF) = 10^{-12} farad, and are sometimes applied to other units (megabit).

6. POWER NOTATION

This expresses concisely any power of ten (any number that is composed of factors of 10). 10^2 or ten squared = $10 \times 10 = 100$; 10^3 or ten cubed = $10 \times 10 \times 10 = 1,000$. Similarly, $10^4 = 10,000$ and $10^{10} = 1$ followed by ten zeros = 10,000,000,000. Proceeding in the opposite direction, dividing by ten and subtracting one from the index, we have $10^2 = 100$, $10^1 = 10$, $10^0 = 1$, $10^{-1} = ¹⁄₁₀$, $10^{-2} = ¹⁄₁₀₀$, and so on; $10^{-10} = ¹⁄₁₀,₀₀₀,₀₀₀,₀₀₀$.

7. BINARY SYSTEM

Only two units (0 and 1) are used, and the position of each unit indicates a power of two.

One to ten written in binary form:

	eights (2^3)	fours (2^2)	twos (2^1)	one
1				1
2			1	0
3			1	1
4	1	0	0	
5		1	0	1
6		1	1	0
7		1	1	1
8	1	0	0	0
9	1	0	0	1
10	1	0	1	0

i.e., ten is written as 1010 ($2^3 + 0 + 2^1 + 0$); one hundred is written as 1100100 ($2^6 + 2^5 + 0 + 0 + 2^2 + 0 + 0$).

Index

THE MAKING SENSE SERIES

Margot Northey with Joan McKibbin
MAKING SENSE
A Student's Guide to Research and Writing
Eighth Edition

Margot Northey, Dianne Draper, and David B. Knight
MAKING SENSE IN GEOGRAPHY AND ENVIRONMENTAL SCIENCES
A Student's Guide to Research and Writing
Sixth Edition

Margot Northey, Lorne Tepperman, and Patrizia Albanese
MAKING SENSE IN THE SOCIAL SCIENCES
A Student's Guide to Research and Writing
Sixth Edition

Margot Northey and Judi Jewinski
MAKING SENSE IN ENGINEERING AND THE TECHNICAL SCIENCES
A Student's Guide to Research and Writing
Fifth Edition

Margot Northey and Patrick von Aderkas
MAKING SENSE IN THE LIFE SCIENCES
A Student's Guide to Research and Writing
Second Edition

Margot Northey, Bradford A. Anderson, and Joel N. Lohr
MAKING SENSE IN RELIGIOUS STUDIES
A Student's Guide to Research and Writing
Second Edition

Margot Northey and Brian Timney
MAKING SENSE IN PSYCHOLOGY
A Student's Guide to Research and Writing
Second Edition